T0134445

Advances in Intelligent Systems and Computing

Volume 913

Series editor

Janusz Kacprzyk, Systems Research Institute, Polish Academy of Sciences, Warsaw, Poland
e-mail: kacprzyk@ibspan.waw.pl

The series "Advances in Intelligent Systems and Computing" contains publications on theory, applications, and design methods of Intelligent Systems and Intelligent Computing. Virtually all disciplines such as engineering, natural sciences, computer and information science, ICT, economics, business, e-commerce, environment, healthcare, life science are covered. The list of topics spans all the areas of modern intelligent systems and computing such as: computational intelligence, soft computing including neural networks, fuzzy systems, evolutionary computing and the fusion of these paradigms, social intelligence, ambient intelligence, computational neuroscience, artificial life, virtual worlds and society, cognitive science and systems, Perception and Vision, DNA and immune based systems, self-organizing and adaptive systems, e-Learning and teaching, human-centered and human-centric computing, recommender systems, intelligent control, robotics and mechatronics including human-machine teaming, knowledge-based paradigms, learning paradigms, machine ethics, intelligent data analysis, knowledge management, intelligent agents, intelligent decision making and support, intelligent network security, trust management, interactive entertainment, Web intelligence and multimedia.

The publications within "Advances in Intelligent Systems and Computing" are primarily proceedings of important conferences, symposia and congresses. They cover significant recent developments in the field, both of a foundational and applicable character. An important characteristic feature of the series is the short publication time and world-wide distribution. This permits a rapid and broad dissemination of research results.

More information about this series at http://www.springer.com/series/11156

Mostafa Ezziyyani
Editor

Advanced Intelligent Systems for Sustainable Development (AI2SD'2018)

Vol 3: Advanced Intelligent Systems Applied to Environment

 Springer

Editor
Mostafa Ezziyyani
Computer Sciences Department,
Faculty of Sciences and Techniques
of Tangier
Abdelmalek Essaâdi University
Souani Tangier, Morocco

ISSN 2194-5357   ISSN 2194-5365 (electronic)
Advances in Intelligent Systems and Computing
ISBN 978-3-030-11880-8   ISBN 978-3-030-11881-5 (eBook)
https://doi.org/10.1007/978-3-030-11881-5

Library of Congress Control Number: 2019930141

This Springer imprint is published by the registered company Springer Nature Switzerland AG
The registered company address is: Gewerbestrasse 11, 6330 Cham, Switzerland

Preface

Overview

The purpose of this volume is to honour myself and all colleagues around the world that we have been able to collaborate closely for extensive research contributions which have enriched the field of Applied Computer Science. Applied Computer Science presents a appropriate research approach for developing a high-level skill that will encourage various researchers with relevant topics from a variety of disciplines, encourage their natural creativity, and prepare them for independent research projects. We think this volume is a testament to the benefits and future possibilities of this kind of collaboration, the framework for which has been put in place.

About the Editor

Prof. Dr. **Mostafa Ezziyyani,** IEEE and ASTF Member, received the "Licence en Informatique" degree, the "Diplôme de Cycle Supérieur en Informatique" degree and the PhD "Doctorat (1)" degree in Information System Engineering, respectively, in 1994, 1996 and 1999, from Mohammed V University in Rabat, Morocco. Also, he received the second PhD degree "Doctorat (2)" in 2006, from Abdelmalek Essaadi University in Distributed Systems and Web Technologies. In 2008, he received a Researcher Professor **Ability Grade. In 2015, he receives a PES grade —the highest degree at Morocco University.** Now he is a Professor of Computer Engineering and Information System in Faculty of Science and Technologies of Abdelmalek Essaadi University since 1996.

His research activities focus on the modelling databases and integration of heterogeneous and distributed systems (with the various developments to the big data, data sciences, data analytics, system decision support, knowledge management, object DB, active DB, multi-system agents, distributed systems and mediation). This research is at the crossroads of databases, artificial intelligence, software engineering and programming.

Professor at Computer Science Department, Member MA laboratory and responsible of the research direction Information Systems and Technologies, he formed a research team that works around this theme and more particularly in the area of integration of heterogeneous systems and decision support systems using WSN as technology for communication.

He received the first WSIS prize 2018 for the Category C7: ICT applications: E-environment, First prize: MtG—ICC in the regional contest IEEE - London UK Project: "World Talk", The qualification to the final (Teachers-Researchers Category): Business Plan Challenger 2015, EVARECH UAE Morocco. Project: «Lavabo Intégré avec Robinet à Circuit Intelligent pour la préservation de l'eau», First prize: Intel Business, Challenge Middle East and North Africa—IBC-MENA. Project: «Système Intelligent Préventif Pour le Contrôle et le Suivie en temps réel des Plantes Médicinale En cours de Croissance (PCS: Plants Control System)», Best Paper: International Conference on Software Engineering and New Technologies ICSENT'2012, Hammamat-Tunis. Paper: «Disaster Emergency System Application Case Study: Flood Disaster».

He has authored three patents: (1) device and learning process of orchestra conducting (e-Orchestra), (2) built-in washbasin with intelligent circuit tap for water preservation. (LIRCI) (3) Device and method for assisting the driving of vehicles for individuals with hearing loss.

He is the editor and coordinator of several projects with Ministry of Higher Education and Scientific Research and others as international project; he has been involved in several collaborative research projects in the context of ERANETMED3/PRIMA/H2020/FP7 framework programmes including project management activities in the topic modelling of distributed information systems reseed to environment, Health, energy and agriculture. The first project aims to propose an adaptive system for flood evacuation. This system gives the best decisions to be taken in this emergency situation to minimize damages. The second project aims to develop a research dynamic process of the itinerary in an events graph for blind and partially signet users. Moreover, he has been the principal investigator and the project manager for several research projects dealing with several topics concerned with his research interests mentioned above.

He was an invited professor for several countries in the world (France, Spain Belgium, Holland, USA and Tunisia). He is member of USA-J1 programme for TCI Morocco Delegation in 2007. He creates strong collaborations with research centres in databases and telecommunications for students' exchange: LIP6, Valencia, Poitier, Boston, Houston, China.

He is the author of more than 100 papers which appeared in refereed specialized journals and symposia. He was also the editor of the book "New Trends in Biomedical Engineering", AEU Publications, 2004. He was a member of the Organizing and the Scientific Committees of several symposia and conferences dealing with topics related to computer sciences, distributed databases and web technology. He has been actively involved in the research community by serving as reviewer for technical, and as an organizer/co-organizer of numerous international and national conferences and workshops. In addition, he served as a programme committee member for international conferences and workshops.

He was responsible for the formation cycle "Maîtrise de Génie Informatique" in the Faculty of Sciences and Technologies in Tangier since 2006. He is responsible too and coordinator of Tow Master "DCESS - Systèmes Informatique pour Management des Entreprise" and "DCESS - Systèmes Informatique pour Management des Enterprise". He is the coordinator of the computer science modules and responsible for the graduation projects and external relations of the Engineers Cycle "Statistique et Informatique Décisionnelle" in Mathematics Department of the Faculty of Sciences and Technologies in Tangier since 2007. He participates also in the Telecommunications Systems DESA/Masters, "Bio-Informatique" Masters and "Qualité des logiciels" Masters in the Faculty of Science in Tetuan since 2002.

He is also the founder and the current chair of the blinds and partially signet people association. His activity interests focus mainly on the software to help the blinds and partially signet people to use the ICT, specifically in Arabic countries. He is the founder of the private centre of training and education in advanced technologies AC-ETAT, in Tangier since 2000.

Mostafa Ezziyyani

Contents

Modeling the Spatial Distribution of Rainfall in the Tangier Area (Northern Morocco)

Boulahfa Imane[1]([✉]), Aboumaria Khadija[1,2], El Halimi Rachid[1,2],
Batmi Abdeladim[2], Maatouk Mustapha[1,2],
and Abattouy Mohammed[1,2]

[1] University Abdelmalik Essaadi,
Faculty of Sciences and Techniques of Tangier, Tangier, Morocco
Imano.boulahfa@gmail.com
[2] Meteorological Center of Tangier Airport, Tangier, Morocco

Abstract. Climate change has become a subject of several studies for our country, especially the far north west of Morocco. This main aims of this work to interpolate the rainfall fields in the Tangier region taking into account the effect of the relief, based on the numerical Terrain model (DEM). To achieve its goals we propose using the method AURELHY (Analyse Utilisant le Relief pour les Bésoins de l'Hydrométéorologie), which allows, from the values of point precipitation their extrapolation to the points not measured Based on Geostatistics, the AURELHY interpolation technique has the advantage of taking into consideration the topography of the region considered according to several stages. The first of these was an integrated approach consists of the coding of the topography surrounding each rainfall station using the Python 3.4 software. The various altitudes are therefore shall be subject by analysis principal components, (ACP) via the R software. In order to identify the dominant trends of slopes for different own vectors. The residues resulting from multiple linear regression between own vectors, longitude, altitude and distance from the sea. Allowed us to make interpolation maps by Kriging of these residues. These intermediate maps were finally used to reconstruct the precipitation fields.

The temporal interval studied was subsequently placed in the context of climate change to identify the impact of this global phenomenon on the reduction of precipitation at the level of the region studied.

Keywords: Climate change · Rainfall field · Interpolation map · Precipitation

1 Introduction

Morocco undergoes the effects of climate change with the specific characteristics of its geographical position and the particularities of its ecosystems. Located in northwestern Africa, between two climatic zones (temperate to the north and tropical to the south), Morocco has a very varied Mediterranean climate: humid, sub-humid, semi-arid, arid and Saharan. The observations of the last decades show the progress of the arid and semi-arid climate towards the north of the country. These changes also include changes in precipitation patterns. These precipitations are dependent, in addition to the global

© Springer Nature Switzerland AG 2019
M. Ezziyani (Ed.): AI2SD 2018, AISC 913, pp. 1–12, 2019.
https://doi.org/10.1007/978-3-030-11881-5_1

parameters, on some very local parameters, such as topography, land-use or atmospheric microphysics (Sebbar et al. 2011).

2 Study Area and Objectives

The region of Tangier is located in the extreme north-west of Morocco between the coordinates (504564, 83N; 589280, 35N) and between (435984, 83W; 523672, 35W). It is characterized by a landscape of average elevation (up to 1300 m) in relation to the sea, it is under the influence of a quite rainy maritime climate, characterized by the gentleness of its temperatures. This region also undergoes the effects of the mountain climate due to its proximity to the mountain range of the Rif. Similarly, during the summer season, the influence of the reliefs is felt by the presence of the Chandler phenomenon caused by the East wind and the effect of Foehn (or Föhn) (Fig. 1).

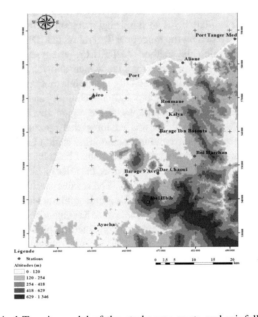

Fig. 1. Numerical Terrain model of the study area posts and rainfall stations used.

In this study we used data from 12 stations and rainfall positions that fall under the direction of national meteorology as well as the direction of hydraulics. The data we used is the normal monthly precipitation of the five years available (2003–2004–2005–2011–2012).

The modeling is done here based on the premise that the closer the data is geographically the more they tend to be similar in value. This modeling has allowed us to achieve spatial predictions and to test hypotheses by explicitly integrating this spatial dependence into the calculations.

3 Material and Methodology

The methodology adopted in this study is based on the principles of statistical modeling. A meteorological file was established after calculating the monthly precipitation from the daily data collected from the rainfall stations in the Tangier region during the study period Fig. 2. A previous processing of this file has been done to extract the desired information.

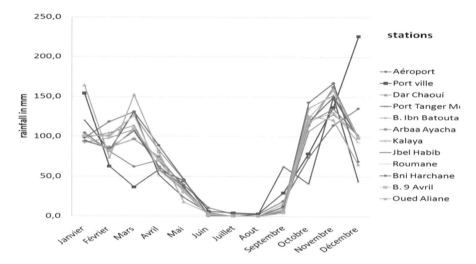

Fig. 2. The twelve stations studied with their monthly precipitation.

3.1 Usual Accumulation of Precipitations

The interpolation methods used in this study have called for a succession of statistical tools requiring a certain stationarity, it seems inappropriate to carry out analyses based on the daily values, these being very Fluctuating. A combination of these values makes it possible to reduce the numerical differences between the different neighbouring posts.

3.2 Application of the AURELHY Method

It is based on the fact that precipitation measured at one point can be broken down into one part in relation to the topography surrounding this point and a part not explained by it. This method is to encode, in vector form, each of the weather stations in the surrounding landscape. This landscape, being defined as a scalar, can be associated with rainfall data (Fig. 3).

After condensed the landscape by a Main-component (PCA) analysis, a regression equation was calculated for the first major components of the landscape at the grid points, and then the residual field of the regression was analyzed by a Kriging method (Benichou et al. 1987).

$$
\begin{array}{c}
site_1 \\
\vdots \\
site_i \\
\vdots \\
site_{12}
\end{array}
\begin{pmatrix}
v_{1,1} & \cdots & v_{1,j} & \cdots & v_{1,441} \\
\vdots & & \vdots & & \vdots \\
& \cdots & G & \cdots & \\
\vdots & & \vdots & & \vdots \\
v_{12,1} & \cdots & v_{12,j} & \cdots & v_{12,441}
\end{pmatrix}
$$

Fig. 3. The matrix of observed sites

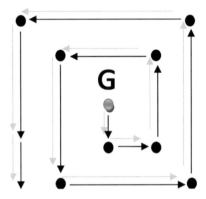

Fig. 4. Order of coding of the landscape surrounding the point G

The components of the model are linked according to the following equation (Fig. 4):

$$observed\ data\ =\ predicted\ data\ +\ interpolation\ of\ residues$$

In a mathematical form, the problem is to find P at the point S of coordinates x, y such as:

$$P(S) \;=\; P(x, y, R) \;=\; \varepsilon(x, y) \tag{1}$$

With:

- P (S): Precipitation at a given point
- R: the landscape vector
- f(R): the regression equation
- $\varepsilon(x, y)$: residues.

The choice of the numerical model of the Terrain to apply the AURELHY method is paramount because the window and the sampling step strongly affect the result.

In this work, we had a 270 m resolution MNT, but since the size of the matrix is large (difficult to manipulate), we have degraded it to 523 m to be able to perform the PCR. So, this case, we chose an 11 km window with a 5.5 km step. The terrain gives an estimated altitude (more or less tainted by errors) at any point in a kilometre grid. The terrain coding is based on terrain altitudes provided by the terrain. This chosen encoding is based on the determination of the "landscape" surrounding a "G" point, considering a square matrix (2n + 1, 2n + 1) of altitude surrounding the point in question.

The dimension 'n' is associated with the assumed radius of influence of the terrain on rainfall which is generally 25 km, but since the study area is not wide enough and given the characteristics determining the microclimate of the region, we can fix this radius to 10 km AFI N to regain the mean-scale variance of the rainfall field.

The landscape surrounding the point is finally a matrix of altitudes of (21 * 21) points. The site "G" is then characterized by 441 values.

The landscape coding (441 points) of a grid point follows a well-defined, sector-organized order to preserve the concept of terrain orientation. The grid of the 441 points is divided from the centre point G while starting with the neighbouring points with variable radius in order to preserve the notion of removal. This encoding was done using the Python programming language.

3.3   Results of the PCA

The principal component Analysis (PCA) was carried out using R software on the altitudes of the various points surrounding the 12 rainfall stations. We selected 6 clean vectors which alone account for a significant percentage of 90% of the variance in the terrain. These clean vectors have interesting physical significance and are: VP 1, VP 2, VP 3, VP 4, VP 5 and VP 6 (Fig. 5).

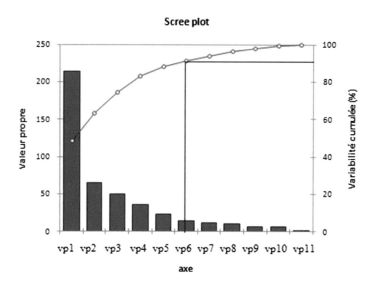

Fig. 5. Eigenvalues with the cumulative percentage of variance explained

3.4   Precipitation Regression with Selected Predictors

The linear regression model (at the level of its parameters) is written:

$$\gamma_i = \beta_0 + \beta_1\alpha_i + \cdots + \beta_0\alpha_i\rho + \varepsilon_i \tag{2}$$

Our study covers the five months of the year in which rainfall is highest: October, November, December, January and February. Through the R software, we realized the multiple linear regressions by adopting the "step-by-Step" selection method (Stepwise) whose selection process begins with the addition of the variable with the highest contribution to the model.

The regression equation developed previously was applied throughout the study area to map the precipitation (Fig. 6).

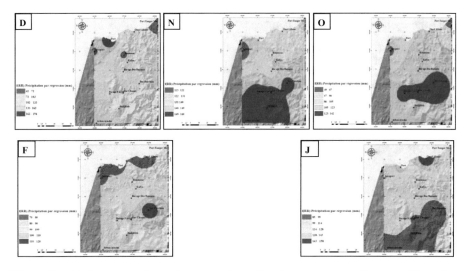

Fig. 6. Precipitaion interpolated by regression, for the months of October, November, December, January and February

3.5   Kriging of Residues

The kriging is part of a process of analyzing geostatistical data. It is necessary to carry out:

- Variographique Analysis
- Adjustment to a theoretical model of the Variogram
- The application of the Kriging

To carry out the analysis of the Variogram, one opted at the beginning for the choice of two types of the kriging: Ordinary and Universal, which one studied for each the different models variographiques (circular, spherical, …) to select the best which adapts to Our study variable. The selection criterion adopted is the minimization of the mean quadratic error (EQM). The study of the different models, allowed us to select the circular Variographique model which is written as follows:

$$h = C0 + C12h\pi a1 - ha2 + 2\pi\, arc\, sin\, ha \qquad (3)$$

The selected parameters of the circular variogram (range, bearing, nugget effect) were used to achieve the kriging of the residues (Fig. 7).

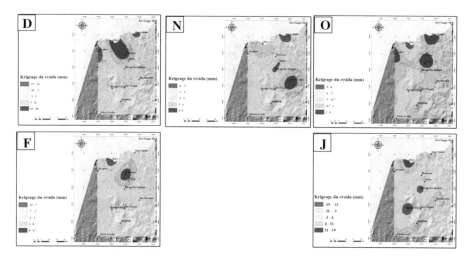

Fig. 7. Precipitation map interpolated by the AURELY method in November period (mm)

4   Results

Precipitation is influenced by several factors such as the soil structure, the type of weather that generates precipitation, wind, vegetation and the topography of rainfall fields. Thus, to improve the quality of precipitation mapping, we have tried to add other predictors (measurable or quantifiable) describing some factors that influence the terrain.

The predictors we have chosen are:

- The terrain's own vectors (VP 1, VP 2, VP 3, VP 4, VP 5 and VP 6)
- The longitude of the station (Y)
- The average altitude of the station (Z)

The average distance from the sea can also be chosen as a predictor, since the region of Tangier lies between two seas (the Atlantic Ocean and the Mediterranean Sea).

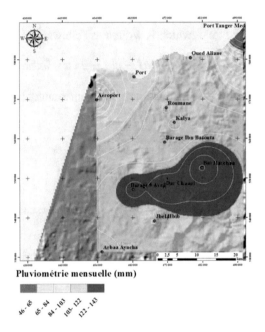

Fig. 8. Precipitation map interpolated by the AURELY method in October period (mm).

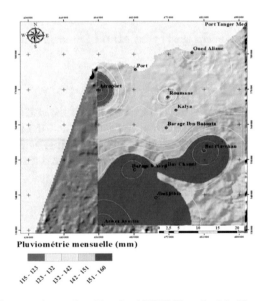

Fig. 9. Precipitation map interpolated by the AURELY method in November period (mm).

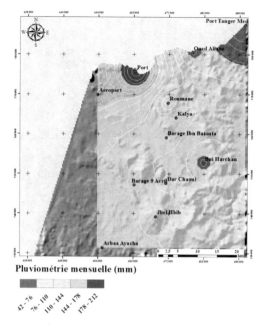

Fig. 10. Precipitation map interpolated by the AURELY method in December period (mm)

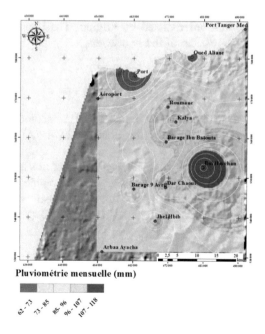

Fig. 11. Precipitation map interpolated by the AURELY method in January period (mm).

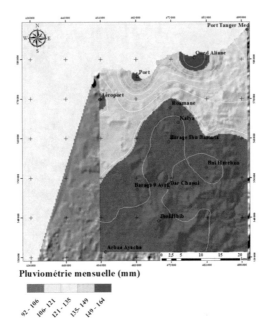

Fig. 12. Precipitation map interpolated by the AURELY method in February period.

To restore the final precipitation field, the map of the regression field is superimposed with the residue field interpolated by kriging. The Figs. 8, 9, 10, 11 and 12 represent the result of the overlap of the two layers.

5 Validation of the Results

In order to verify the results provided by the AURELHY method, a comparison of the precipitation normals interpolated by kriging alone and by the AURELHY method was carried out.

It is noted that the AURELHY method has given more precision, thus fewer errors than the kriging alone, proving that the addition of topographic information improves the interpolation of the precipitation field. This fact is easily confirmed when comparing the maps obtained by the method AURELHY and those by kriging of the normal precipitation.

6 The Impact of Climatic Warming on Rainfall Intensity

Significant climatic changes affect a large part of Morocco in general. This study shows particularly the climate change impacts in the Tangier region. It highlighted the spatio-temporal evolution of monthly and annual precipitations in the study area. On the other hand, it has been confirmed that, from the 1970s, annual and monthly rainfall accumulations have undergone significant downward trends.

This study clearly discerns a general trend towards a decrease in precipitation which confirms our results obtained.

Taking the example of the month January in the station of Tangier Airport and based on the (Table 1).

Table 1. Normal monthly precipitation in (mm).

Climatological normals	October	November	December	January	February
1971–2000	73,9	114,5	135,9	104,5	81,6
1981–2010	80,7	127	133,9	94,1	76,7

Table 2. Normal annual rainfall in (mm)

Annual normals from 1971–2000	696.6
Annual normals from 1981–2010	717.1

(Source direction of national meteorology airport Tangier Ibn Batouta)

It is noted that January tends towards a 10% decrease in rainfall compared to the two climatological normals, this shows the impact of climate change on rainfall intensity in the Tangier region. The annual climatic cumulation also has a trend towards a 3% decrease in rainfall, according to the (Table 2).

However, the analysis of rainfall fields for short-term accumulations still poses many problems. However, the point values that we have been able to achieve remain very fluctuating, since we have only been able to establish them over a period of five years.

In order to achieve effective and more operational results, it would be ideal to extend the study over a period of thirty years and carry it out on an even wider area with a larger number of stations that would cover as much space as possible.

7 Conclusion

The climate is warming and the consequences on the planet are likely to be disastrous. such as water scarcity, precipitation reduction…

As we have seen, the assessment of the mapping of rainfall fields, based on Geostatistic, is a complex approach, but which allows to provide estimated values at any point depending on the terrain.

References

Arnaud, M., Emery, X.: Spatial Estimation and Interpolation, Deterministic Methods and Geostatistical Methods. HERMES Science Europe, 219 p. (2000)

Benichou, P., Le Breton, O.: Taking into account topography for the mapping of statistical rainfall Fields, Directorate of National Meteorology of France (1986)

El Khatri, S.: Analysis of multidimensional data. Directorate of National Meteorology and Ecole Hassana of public works Casablanca, Morocco, December 2004

Gratton, Y.: The kriging: the optimal method of spatial interpolation 2002

Intergovernmental Panel on Climate Change, 2007: Review of climate change: physical scientific elements. In: Solomon, S., Qin, D., Manning, M., Chen, Z., Marquis, M., Averyt, K.B. et al. (eds.) Fourth Assessment Report of the Intergovernmental Panel on Climate Change. Cambridge University Press, Cambridge (2007)

Sebbar, A., Badri, W., Fougrach, H., Hsain, M., Saloui, A.: Study of the variability of the rainfall regime in Northern Morocco (1935–2004). Drought **22**(3), 139–148 (2011)

Marcotte, D.: Mining and Cokriging geostatistical course with MATLAB (1991). http://geo.polymtl.ca/

Author, F.: Article title. Journal **2**(5), 99–110 (2016)

Author, F., Author, S.: Title of a proceedings paper. In: Editor, F., Editor, S. (eds.) CONFERENCE 2016, LNCS, vol. 9999, pp. 1–13. Springer, Heidelberg (2016)

Author, F., Author, S., Author, T.: Book Title, 2nd edn. Publisher, Location (1999)

Author, F.: Contribution title. In: 9th International Proceedings on Proceedings, pp. 1–2. Publisher, Location (2010)

LNCS Homepage. http://www.springer.com/lncs. Accessed 21 Nov 2016

Comparison of the Relevance and the Performance of Filling in Gaps Methods in Climate Datasets

Jada El Kasri[✉], Abdelaziz Lahmili, Ouadif Latifa, Lahcen Bahi, and Halima Soussi

3GIE Laboratory, Mineral Engineering Department, Mohammadia Engineering School, Mohammed V University, Rabat, Morocco
jada.elkasri@gmail.com

Abstract. The lack of values in a climatological series is a severe problem that can mislead and mistake scientific studies. The purpose of this study is to compare three methods of filling in the missing data; the simple arithmetic averaging (AA), Inverse distance interpolation (ID) and the multiple imputation (MI). The comparison of these methods was carried out on a list of mean monthly temperature that concerns one hydrological station localized in the basin of Souss, and was based on four evaluation criteria, namely root mean square error (RMSE), mean absolute errors (MAE), skill score (SS) and coefficient of efficiency (CE). The analysis shows the effectiveness of multiple imputation and the application of the performance criteria shows that MI had the lowest error measures, the best coefficient of efficiency and the best Skill Score. Therefore, we recommend the use of MI to resolve the gap in climatic datasets, especially large ones.

Keywords: Climate datasets · MI · Missing data · MAR

1 Introduction

Missing values in datasets is an old and common problem in most scientific research and in different domains. That might arise from various sources depending on the field. In the area of climate science, missing values in climatic datasets are frequently encountered; they are due mostly to a failure of measuring instruments of the observatory, a recurring breakdown of communication line, or because of the absence of the observer [1]. For a good interpretation and an effective analysis, climatic studies need complete data which are recorded in many stations situated over the region under the study, and yet sometimes researchers face a serious problem, missingness of values

© Springer Nature Switzerland AG 2019
M. Ezziyyani (Ed.): AI2SD 2018, AISC 913, pp. 13–21, 2019.
https://doi.org/10.1007/978-3-030-11881-5_2

might show up in other nearby stations, and that can gravely hinder the results of the study. For the success of the study, it is crucial to choose the appropriate method to fill in gaps, since ignoring the missing data can lead to a significant bias in the analysis models and a loss of precision. For this reason, many statistical and empirical techniques have been used by scientifcs to estimate missing data, such as simple arithmetic mean (MAS), inverse distance interpolation (ID) [2]. Concerning statistical methods, there are the multiple regression analysis (REG) [2] and multiple imputation [3] which are mostly used. MI is proved to be a successful method for dealing with missing data, and it is becoming increasingly common in many fields, though this method is still relatively rarely used in climatic studies.

In order to properly succeed the imputation of missing data, it is inevitable to distinguish the causes of disparity. A typology has been developed by Little and Rubin in 2002 [4], tabulating mechanisms of disparity into three categories namely missing at random data MAR in this assumption; the missing variable depends only on observed variables. Missing completely at random data MCAR: A variable is MCAR if the probability of missingness is independent of any characteristics of the subjects. Missing not at random mechanism MNAR: The missingness is no longer "at random" if its probability depends on variables that are incomplete. [5] Schneider proves that, temperature data are included in missing at random assumption because the holes that appear in temperature series does not depend absolutely on the temperature itself but may depend on other variables [6]. The same reasoning can be applied on other climatic variables, such as temperature and relative humidity.

This paper focuses on a situation in which a large dataset, of monthly mean, contains a lot of gaps, and it is hard to use data from adjacent stations for diverse calculations and estimations. It also investigates the validity of three methods: the simple arithmetic averaging (AA), Inverse distance interpolation (ID) and the multiple imputation (MI). Finally, to deduct the most useful method in such situations in order to help researchers solve the problem.

2   Data Description

The station "Taroudant-bridge" under study is located in the basin of Souss situated in the western zone of southern Atlas of Morocco. Dataset used in the study is the mean monthly temperature measured over 16 years from 1982 to 1998. Figure 1.

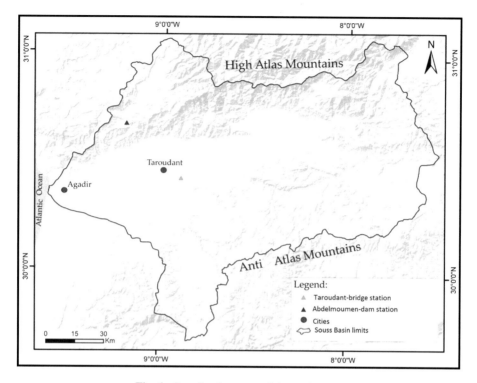

Fig. 1. Localization map of the study area

3 Materials and Methods

In the station "Taroudant-bridge", some of the observed data have been purposely deleted in order to predict them subsequently by using the three methods. The simulated missing values were imputed using the following methods: Simple arithmetic averaging (AA), Inverse distance interpolation (ID) and multiple imputation (MI). Station Abdelmoumen-dam is the only station that has recorded data on the needed duration and its measures are used in calculations. For the soudness of the imputed values, four evaluation criteria (RMSE, MAE, SS and CE) were used. The following flowchart presents the methodology used in the analysis (Fig. 2).

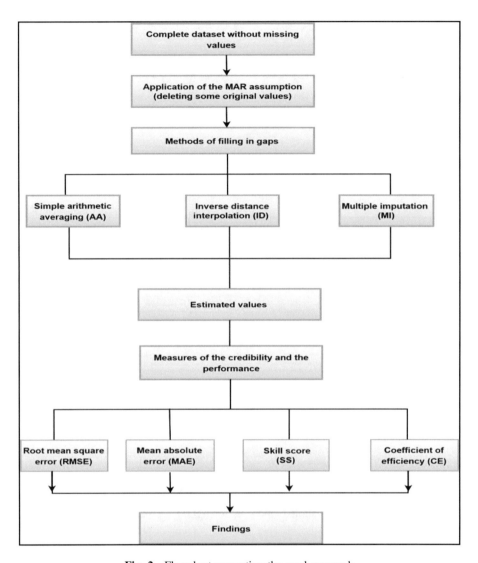

Fig. 2. Flowchart presenting the used approach

3.1 Arithmetic Averaging (AA)

Arithmetic averaging is a simple method of calculating missing climate data over a basin. The result is obtained by the division of the sum of the monthly temperature data recorded at nearest stations of the basin by the number of the stations. The equation would be as follows:

$$T_0 = \frac{\sum_{i=1}^{n} T_i}{n} \tag{1}$$

T_0 is the estimated value of the missing data. T_i is the value of the same parameter at ith nearest hydrological station on a given date, and n is the number of the nearest stations.

3.2   Inverse Distance Interpolation (ID)

The inverse distance method is used for its simplicity to implement missing data [7]. This method takes into consideration the shortest distance between two different stations therefore, the equation would be as:

$$T_0 = \frac{\sum_{i=1}^{n} \left(\frac{T_i}{d_i}\right)}{\sum_{i=1}^{n} \left(\frac{1}{T_i}\right)} \tag{2}$$

T_0 is the estimated value of the missing data. T_i is the value of the same parameter at ith nearest hydrological station on a given date. d_i is the distance between the surrounding stations and the station in question, and n is the number of the nearest stations.

3.3   Multiple Imputation

The multiple imputation (MI) is a method developed by Rubin in 1987. It should be applied strictly when the missingness occurs under a missing at random data assumption. The objective of this method is to perform multiple imputations usually repeated five times on the same set of data. Each set of imputed data is, then, analyzed separately. The end result is the average of the results of the analyses [8, 9]. It is available in programs most commonly used in studies by statisticians, such as SPSS [10].

3.4   Performance Criteria

The Performance criteria enable us to measure the performance of the imputation methods at estimating point values.

Root Mean Square Error (RMSE)
Root mean square error (RMSE) is a measure used to reveal the difference between the predicted values by a model and the real observed values. The model that gives the lowest value of RMSE is indicated to be a valid and reliable model. The range of RMSE is from 0 to $+\infty$. [11, 12].

It is calculated by this formula:

$$RMSE = \sqrt{\frac{\sum_{i=1}^{n}(T_{obsi} - T_{esti})^2}{n}}$$

Mean Absolute Errors (MAE)

The average absolute error (MAE) measures the amount of the estimate error. Willmott and Robeson suggested its use [13]. The method that gives the lowest value for MAE is considered to be a valid and reliable method of prediction. As like RMSE the range of MAE is also from 0 to $+\infty$ [14].

The formula is:

$$MAE = \frac{\sum_{i=1}^{n}|T_{obsi} - T_{esti}|}{n}$$

Coefficient of Efficiency (CE)

It indicates the power of the regression model. It ranges from $-\infty$ to 1. The higher the coefficient of efficiency is the more accurate the model is [14].

$$CE = 1 - \frac{\sum_{i=1}^{n}(T_{obsi} - T_{esti})^2}{\sum_{i=1}^{n}(T_{obsi} - T_{ave})^2}$$

Skill Score (SS)

It is a quality verification measure of a predicted model. When SS = 1 it means a perfect forecast. The closer the SS to 1 is the more reliable the method becomes.

$$SS = 1 - \frac{\sum_{i=1}^{n}(T_{esti} - T_{obsi})^2}{\sum_{i=1}^{n}(T_{ave} - T_{obsi})^2}$$

4 Results and Discussion

According to results, the three methods performed well in the sense that they do not show much variation and have quiet close values of the errors measures coefficient of efficiency and skill score. In fact, in Figs. 3 and 4, they all show low values of RMSE and MAE. It must be noted that MI has the lowest value of RMSE, but it takes the highest value of MAE. On Fig. 5, the MI has the best skill score and the highest coefficient of efficiency which allow us to decide the best method of imputation for this case. AA and ID are still valid methods. However, in our study we select the best method which can impute estimated values that can almost match real observed values.

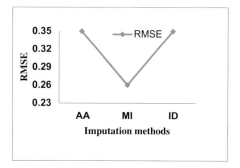

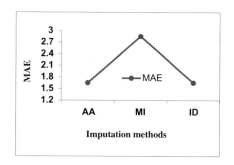

Fig. 3. Presentation of the RMSE of each method

Fig. 4. Presentation of the MAE of each method

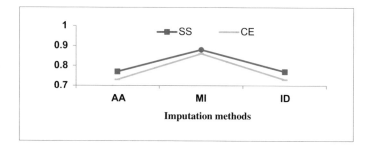

Fig. 5. Presentation of the SS and CE of each method

Based on the results, multiple imputation MI, performed with SPSS software, is the most suitable method for estimating missing monthly temperature data. Arithmetic averaging (AA) is a commonly used method that gives reliable results, but only when stations in the basin area are uniformly distributed and the values do not vary widely [15]. Inverse distance interpolation (ID) is a good imputation method because it takes into consideration the weight of the stations. In our case, we did the calculations of AA and ID using the measures of the station "Abdelmoumen-dam" considered as the nearest station without missing values, since it is the only station: the estimated values are as the same as the station Abdelmoumen's values. Multiple imputation has proved its validity to resolve different datasets of missingness issues. It is known to be applicable for big datasets [16]. In another study, we applied multiple imputation on a small dataset of annual mean rainfall and its effectiveness was acceptable and we confirmed that for climatic data, MI should be applied on large samples [17]. It is confirmed to be a versatile method, adaptable to different seasons, and should be considered as an alternative to fill in gaps in time series of precipitation [18].

5 Conclusion

In a world where information is power, missing values in datasets continue to pose a serious hindrance. Climate studies still suffer from this phenomenon for a variety of reasons. In this study, we presented three practical methods in order to fill in gaps in monthly temperature datasets that are AA, ID arithmetic methods and MI a statistical method incorporated in SPSS statistical software. The three methods of estimating missing monthly temperature data were applied on the station of "Taroudant-bridge" and their effectiveness was evaluated using four skill criteria, namely root mean square error (RMSE), mean absolute error (MAE), coefficient of efficiency (CE) and skill score (SS). After analysing the results, we may recommend multiple imputation as the best method for filling in gaps of missing monthly mean temperature data in large samples.

References

1. Aslan, S., Yozgatligil, C., Iyigun, C., Batmaz, I., Turkes, M., Tatli, H., Batmaz, I.: Comparison of missing value imputation methods for turkish monthly total precipitation data (2014)
2. Xia, Y., Fabian, P., Stohl, A., Winterhalter, M.: Forest climatology: estimation of missing values for Bavaria, Germany. Lehrstuhl fuÈr Bioklimatologie und Immissionsforschung, Ludwig-Maximilians UniversitaÈt MuÈnchen, Am Hochanger 13, 85354 Freising, Germany Received 25 September 1998; received in revised form 11 March 1999; accepted 23 March 1999
3. Yuan, Y.C.: Multiple Imputation for Missing Data: Concepts and New Development P267-25. SAS Institute Inc., Rockville, 1700 Rockville Pike, Suite 600, Rockville, MD 20852
4. Little, R.J.A., Rubin, D.B.: Statistical Analysis with Missing Data, 2nd edn. Wiley-Interscience, New York (2002)
5. He, Y.: Missing data analysis using multiple imputation: getting to the heart of the matter. Circ. Cardiovasc. Qual. Outcomes 3(1), 98–105 (2010)
6. Schneider, T.: Analysis of incomplete climate data: estimation of mean values and covariance matrices and imputation of missing values. J. Clim. **14**, 853–871 (2001)
7. Hubbard, K.G.: Spatial variability of daily weather variables in the high plains of the USA. Agric. For. Meteorol. **68**, 29–41 (1994)
8. Little, R.J.A., Rubin, D.B.: Statistical Analysis with Missing Data. Wiley, New York (1987)
9. Schafer, J.L.: Multiple imputation: a primer. Stat. Methods Med. Res. **8**, 3–15 (1999)
10. Teresa, A.M.: Goodbye, listwise deletion: presenting hot deck imputation as an easy and effective tool for handling missing data. Commun. Methods Measur. **5**(4), 297–310 (2011)

11. Chai, T., Kim, H.-C., Lee, P., Tong, D., Pan, L., Tang, Y., Huang, J., McQueen, J., Tsidulko, M., Stajner, I.: Evaluation of the United States National air quality forecast capability experimental real-time predictions in 2010 using air quality system ozone and NO2 measurements. Geosci. Model Dev. **6**, 1831–1850 (2013)
12. Chai, T., Draxler, R.R.: Root mean square error (RMSE) or mean absolute error (MAE)? – Arguments against avoiding RMSE in the literature. Geosci. Model Dev. **7**, 1247–1250 (2014)
13. Willmott, C.J., Matsuura, K., Robeson, S.M.: Ambiguities inherent in sums-of squares-based error statistics. Atmos. Environ. **43**, 749–752 (2009)
14. Kashani, M.H., Dinpashoh, Y.: Evaluation of efficiency of different estimation methods for missing climatological data. Stoch. Environ. Res. Risk Assess. **26**, 59–71 (2012)
15. Bhavani, R.: Comparision of mean and weighted annual rainfall in anantapuram district. Int. J. Innovative Res. Sci. Eng. Technol. **2**(7), 2794–2800 (2013)
16. Sunni, A.B., Stacy, R.L., Seaman Jr., W.J.: Multiple Imputation Techniques in Small Sample Clinical Trials. Wiley InterScience, Hoboken (2005)
17. El kasri, J., Lahmili, A., Ouadif, L., Bahi, L., Soussi, H., Mitach, M.A.: Comparison of the relevance and performance of filling in gaps methods in rainfall datasets. Int. J. Civil Eng. Technol. (IJCIET) **9**(5), 992–1000 (2018). Article ID: IJCIET_09_05_110
18. Carvalho, J.R.P., Monteiro, J.E.B.A., Nakai, A.M., Assad, D.E.: Model for multiple imputation to estimate daily rainfall data and filling of faults. Revista Brasileira de Meteorologia **32**(4), 575–583 (2017)

Hydrogeological and Hydrochemical Study of Underground Waters of the Tablecloth in the Vicinity of the Controlled City Dump Mohammedia (Morocco)

J. Mabrouki[1]($\boxtimes$), A. El Yadini[1], I. Bencheikh[1], K. Azoulay[1],
A. Moufti[2], and S. El Hajjaji[1]

[1] Laboratory of Spectroscopy, Molecular Modeling, Materials, Nanomaterial,
Water and Environment, CERNE2D, Faculty of Science,
Mohammed V University in Rabat, Avenue Ibn Battouta,
BP1014 Agdal, Rabat, Morocco
jamalmabrouki@gmail.com
[2] Regional Center for Careers in Education and Training,
Casablanca, Settat, Morocco

Abstract. The changes in the wetlands are very dynamic and they have been evaluated from different data using several techniques (comparison of historical maps, comparison of topographic maps, photo-interpretation …).

The BeniYakhlef plain is highly threatened by water pollution related to chemicals, wastewater, solid waste discharges and intensive use of fertilizers. It is useful to monitor the quality of water resources. This monitoring needs the realization of the vulnerability map to groundwater pollution while relying on digital terrain model data and exogenous data such as: geological, climatic, soil and data on aquifers. Using the hydrological balance equation and Geographical Information Systems (GIS), map of the volumes run-off and infiltrated into the watersheds of the BeniYakhlef plain and the estimation of the impact of urban sprawl on water from the study area were done.

The objective of this work is to evaluate the influence of the flow on the physicochemical quality the study site's well's water and to determine the shape of the water table near the water discharge. The results of measurements of the electrical conductivity of groundwater and those of physicochemical analyzes made it possible to define different hydrogeochemical domains. The waters are all on facies mineralized in calcium and magnesium bicarbonates in general. The groundwater flow direction of the water table is plio-quaternary from SW to NE.

Keywords: Groundwater · Hydrodynamisme · Hydrochemical ·
Controlled discharge · Mohammedia · GIS · Hydrogeology

1 Introduction

Groundwater plays a fundamental role in the stability of rural populations in the BeniYakhlef region, Mohammedia, Morocco. They are exploited by wells, springs and boreholes; drained by various traditional and modern techniques used to extract groundwater for drinking water supply and irrigation.

© Springer Nature Switzerland AG 2019
M. Ezziyyani (Ed.): AI2SD 2018, AISC 913, pp. 22–33, 2019.
https://doi.org/10.1007/978-3-030-11881-5_3

The piezometric study of the superficial aquifer made it possible to specify the direction of groundwater flow that is generally from east to west. It also shows a supply of this sheet by the limestone massif, as well as by the direct infiltration of rainwater [1]. The delimitation of the watershed and the extraction of the hydrographic network as well as the realization of the slope map by using the DEM (digital elevation models) images [2], were able to determine the drainage of surface water and their path in the basin.

The methodology used for this work has the particularity of distinguishing the vertical vulnerability of a water table to a pollution coming from the surface of the ground, in the region of Mohammedia. The qualitative study of groundwater was based on graphical illustration (Piper diagrams) and cartographic (hydrochemical maps). The groundwater of the Beni Yakhlef plain reveals one of the chlorinated and sulphated chemical facies due to the geology of the aquifer and the surface and time of water-rock interaction [3, 4]. Despite this chemical diversity, water is generally suitable for human consumption, with the exception of wells located near farms. Also for irrigation [5, 6], water present a general good quality.

2 Situation and Description of the Site

2.1 Location of the Study Area

The Mohammedia landfill is based on the rational storage of solid waste in order to avoid any risk of harm to human health and the environment. It is located on a site of rather impermeable lithological nature, and designed to take care of five large bins, which are dug as and when needed, controlled, constituted of drainage network and collection, allowing recovering and treating the leachate formed in the landfill respecting the fixed values relating to abstraction and water consumption [7].

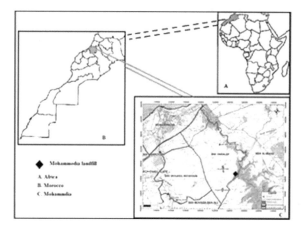

Fig. 1. Map of Morocco and location of the landfill site [8].

The technical center of embankment located on the road P3313, municipality of Chaâba El Hamra Ru-rale of BeniYakhlef, Mohammedia, Morocco, hosts since 27 February 2012 household waste and similar waste from local authorities and companies.

2.2   Geology of the Study Area

The section made at Chaâba El Hamra (Fig. 2) shows from the base to the summit:

- Micaceous schists of the Cambro-Ordovician, covered in angular discordance by the Triassic formations.
- Silto-greso-conglomeratique ensemble of about 20 m of depth, it contains the iron mineralization of Chaâba El Hamra [9–11]. They consist of polygenic conglomerates, subterranean hematitized siltstones and sandstones.
- Lower siltstones and argillites: are reddish with a depth of about 40 m probably Triassic age [12].

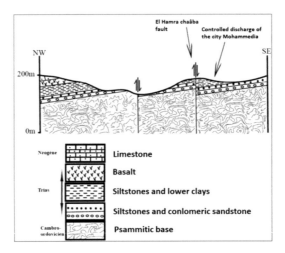

Fig. 2. Geological sections in the landfill sector [11].

- Suctions of vacuolar basalts very weathered (<20 m) of gray-green color on patina. The vacuoles are filled late with quartz and calcite.
- Whitish limestones of probably Neogene age.

3   Materials and Methods

3.1   Piezometric Level Measurement

In hydrogeology, and in the piezometric, the working tools are varied and more precise than the others. These tools allow an accurate and complete study based on measurements of latitudes, longitudes, altitudes and piezometric levels [13].

Piezometric level measurement is the main operation in the groundwater resource inventory. It corresponds to the altitude of the water level in natural balance (sounding or well). It is calculated by difference between the soil dimension (mark on the structure), the altitude Z and the depth of the water [14].

The static levels of each structure are measured using a piezometric probe. This consists of a probe attached to a conductive metric tape. When in contact with water, the probe emits an audible alarm, which allows us, thanks to the metric tape, to measure the height of the surface of the water table with respect to the surface of the ground, or a mark, whose height compared to the ground, is known. The equipment used to accomplish this field work is the Global Positioning System (GPS) for positioning the water points. The data processing is done by the software ARCGIS 10.3.

Software Surfer allows to create Grids that will interpolate the irregular data of our points x, y, z so the ordinates.

3.2   Materials and Methods in Hydrochemistry

In hydrochemistry, the aim is to determine the physicochemical parameters of the groundwater of the wells around the landfill. Seventeen water samples are taken in plastic bottles from the wells trapping the region sheet (Fig. 1). Samples were taken during December 2017 [15]. Water sampling was chosen in distribution zones in order to cover the entire study area [16].

Some measurements were carried out on the spot in order to be able to determine quantitatively the parameters that fluctuate after sampling, such as temperature, pH, electrical conductivity and dissolved oxygen according to the NF EN ISO 19458 standard [17].

4   Results and Discussion

4.1   Piezometry of the Plio-Quaternary Aquifer

The depth of the water table in the study area varies from less than 60 m in the feeding areas near the faults, and more than 40 m in the center, or even more than 65 m near the fault of Chaâba El Hamra (Table 1). Underground flow is controlled by transverse and longitudinal faults (Fig. 3).

The piezometric curve shows many anomalies, namely depression (elliptic curve with converging currents) towards Oued N'fifikh that represents a zone of loss that means that the latter is fed by the water table.

In addition to the depressions, there is the presence of half a tenth of protuberances (elliptic curve with divergent current liners: dome piezometry) encircling the discharge site corresponding in general to large feed sites from the surface (feed through precipitation water spills surface water by fault).

The map presented isopieze curve elliptical cone-shaped depressed with a variable spacing that allows us to characterize the Convergent Radial or Convex-type sheet whose concavity is oriented downstream.

Table 1. Well data in the study area

Id	Depth (m)	Nival (m)	Z(m)	X (m)	Y(m)
0	66	62	152	323226,848211	337876,978508
1	49	38	164	322963,322684	337960,587009
2	37	34	141	32737,230482	336907,80782
3	23	13	154	324554,265449	336958,211046
4	19	0	133	324916,115631	337531,267942
5	18	12	161	325072,654701	336924,209777
6	18	13	162	325334,746036	337151,492107
7	42	34	179	325756,549279	336934,44772
8	26	24	171	326009,09085	337405,519017
9	75	54	148	323112,888499	339103,417201
10	53	51	134	322082,236802	340273,754036
11	47	41	158	324406,773258	340661,197559
12	35	33	149	325340,087751	341053,499615
13	5	0	37	325523,82564	340196,066545
14	7	0	40	325520,833934	339420,703669
15	8	0	47	325231,838814	338276,842615
16	10	0	57	326469,526314	337900,155115

In this case, the flow section decreases in the direction of flow. This type of aquifer characterizes the general drainage zones of the groundwater either on the surface towards the watercourse or at depth.

4.2 Hydrodynamisme of Aquifer

As part of this study, companions in groundwater measurements and sampling were conducted in December 2017. This is to prepare piezometric maps representative of the current state of the groundwater levels of the aquifer. It should be noted that there is heterogeneity in the spatial distribution of measurement points simply due to the geographical location of the different wells.

Using these different measurements, the piezometric water map was produced (Fig. 3). These maps allowed us to determine the overall direction of flow in the aquifer system.

Figure 5 shows the variation of the level of water in the aquifer at wells (P1, P2… P16) (situation of the wells: see Fig. 4). This leads us to conclude that the variation of the piezometric surface of the water table in the region is dependent on the variation of the precipitation and the lateral feedings of the groundwater. So, we see that the water level is high in the works, near the feeding areas and on everything next to faults.

The water level rises with the return of precipitation and the direction of groundwater flow is to the North. The flow of the web can be influenced by the geometry in tilted blocks of its substratum, which causes the deepening of the mio-pliocene raincoat and the thickening of the aquifer (Fig. 6).

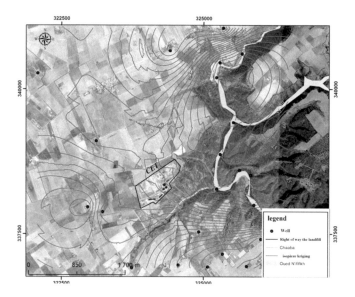

Fig. 3. Piezometric map of study area.

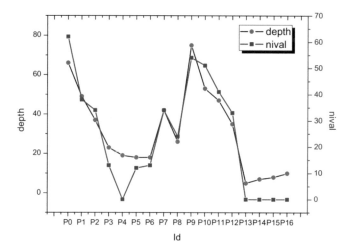

Fig. 4. Fluctuations of the piezometric surface of the aquifer wells

Its interpretation translates a flow of the water table to the north and to the west. There are two watersheds in this coastal zone: the first in the southwest and the second in the east.

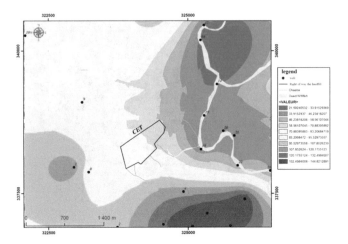

Fig. 5. Fluctuations of the piezometric surface of the aquifer wells

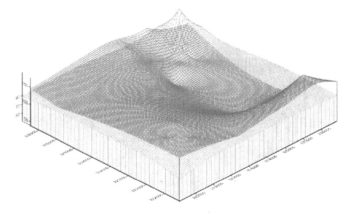

Fig. 6. Superposition 3D des cotés piézométriques et cotés TN des puits de mesure

4.3 Watershed of the Study Area

The hydrographic network refers to the geographical and physical characteristics of the plain and rivers of a given region. The hydrographic network in the plain of Chaâba el Hamra is very developed, with a stream draining this watershed, of which more than half are intermittent streams. The N'fifikh wadi river is very long (Fig. 7).

The depth of the water table represents the vertical distance traversed by a contaminant at the surface of the ground to reach the water table. Generally, the potential protection of the aquifer increases with the depth of the aquifer [18].

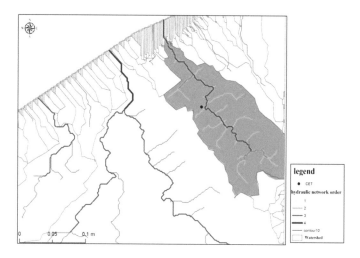

Fig. 7. Hydrographic network map

4.4   Hydrogeochemistry

Measurements of chemical parameters show that the groundwater of the aquifer of the study area is characterized by the abundance of Na^+, Cl^-, SO_4^{2-} and Ca^{2+} ions (Table 2).

Table 2. Results of physicochemical parameters

Id	P0	P1	P2	P3	P5	P6	P7	P8	P9	P10	P11	P12
Cond. (µs/cm)	2430	3245	2741	2024	3112	3045	4562	3014	2450	2157	2653	2254
Ph	7.3	7.14	7.54	7.3	7.6	7.24	7.64	7.6	7.35	7.5	7.4	7.4
T(c°)	13	14	14	15	15	16	14	14	13.4	15.4	16.2	15
O_2d (mg/l)	6.6	7.4	7.80	8.4	8.1	8.35	7.42	9	6.5	8	8.5	8.5
Ca^{2+} (mg/l)	142	154	136	154	142	146	164	150	165	164	125	125
Mg^{2+} (mg/l)	52	41	65.5	58	62.4	62.4	70	61	54	54	54	53
Na^+(mg/l)	352	425	426	425	450	457	502	503	510	520	220	243
K^+(mg/l)	2.5	2.14	1.84	2.4	2.04	2.48	2.14	2.1	2.2	3.5	4.4	3
HCO_3^- (mg/l)	26	30	24	41	34	30	26	42	25	24	45	28
Cl^-(mg/l)	532	532	694	542	654	742	764	642	640	542	642	530
SO_4^{2-}(mg/l)	124	123	134	154	84	92.2	152	154	120	82	72	74
NO_3^- (mg/l)	71	60	64	64	26	25.1	66	56	54	64	64.2	63.1
NO_2^- (mg/l)	0.012	0.012	0.065	0.012	0.024	0.024	0.083	0.012	0.012	0.012	0.012	0.014

The report on the results of the chemical analyses of the aquifer waters on the triangular diagram of Piper (Fig. 8) highlights the impact of lithology on the quality of groundwater. The diagram shows two families of water: the dominant calcic and magnesian Bicarbonated waters and the chlorinated, sulphated, calcic and magnesian waters [19].

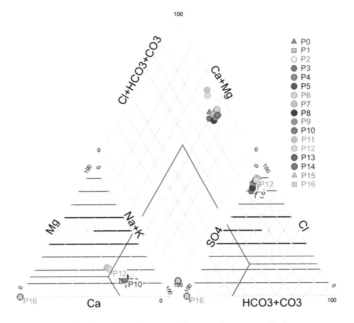

Fig. 8. Piper diagram of groundwater studied

The obtained results in the study area show that the waters of the plain of Chaâba El Hamra are dextrement salty, with very high conductivity. The pH value varies between 7.1 and 7.65. The temperature is between 13.2 and 33.5 °C. The chemical facies of these waters are sulphated chlorate considering the high concentrations in chloride and sulphates [20]. The interpretation of the evolution of nitrate levels shows a degradation of the natural state of groundwater which indicates the pollution of this water [21] (Fig. 9).

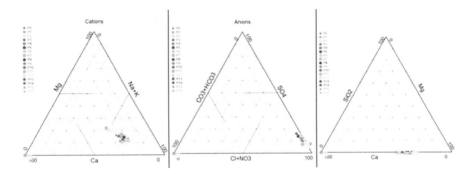

Fig. 9. Piper diagram of anions and cations.

The degradation is linked to several factors; the most important is agricultural activity where the excessive use of fertilizers contributes to the increase of nitrate levels. The waters of irrigation and rain play a major role in the transport of nitrate ions from the soil to the aquifers. Despite this chemical diversity, the waters are generally fit for human consumption in some wells, for those that are close to farms and for irrigation as well.

During the value chain of the project, each activity in the landfill site may affect the various compartments of the environment, particularly the quality of groundwater through several ways that have a major impact:

– Infiltration of leachates through a breach in the geomembrane or following the overflow of storage basins.
– Details of oils, lubricants and hydrocarbons.
– Leachate treated in natural receptacles without reaching the relevant standards.

Any time the laboratory analysis has proved that the hydraulic permeability in site of 10–15, so the time necessary to reach the water table in case of spill would be intoxicating Billions of years. The run off of water in is a means of spreading many pollutants. This is likely to have a major impact on the groundwater as the downstream of the landfill.

5 Conclusion

The flow of groundwater in the fluvio-lacustrine and volcanic deposits is SW to NE towards the sea. This widespread movement of groundwater is guided by collapse and tectonic that affect the region.

The evolution of piezometry over time shows seasonal fluctuations in relation to precipitation. The relationships between the Na^+, Cl^-, SO_4^{2-}, Ca^{2+} and HCO_3-; chemical elements that characterize the geological formations of the region, show the dominance of carbonate ions (Ca^{2+}, HCO_3-) with respect to ions (Ca^{2+}, SO_4^{2-}) and to saline ions (Na^+, Cl^-). The geological context and the spatial distribution of the chemical elements indicate that the chemical composition of the groundwater of the aquifer is strongly influenced by the carbonate's dissolution, clay-evaporitic striasic formations and by the hydrological parameters of the region; namely the direction of flow and the residence time in the aquifer of the plateau.

Acknowledgment. The corresponding author would like to thank The International Conference on Advanced Intelligent Systems for Sustainable Development applied to Agriculture, Energy, Health and Environment for the support given to participate at the conference, 12–14 July 2018, Tangier, Morocco.

References

1. Aller, L., Lehr, J., Petty, H., Bennett, T.: DRASTIC: a standardized system to evaluate groundwater pollution using hydrogeologic setting. J. Geol. Soc. India **29**(1), 23–37 (1987)
2. Castany, G.: Principes et méthodes de l'hydrogéologie, p. 238. Bordas, Paris (1982)
3. Frances, A., Paralta, E., Fernandes, J., Ribeiro, L.: Development and application in the Alentejo region of a method to assess the vulnerability of groundwater to diffuse agricultural pollution: the susceptibility index. In: FGR 2001, Third International Conference on Future Groundwater Resources at Risk, pp. 35–44. CVRM publ., Lisbon (2002)
4. Fedan, B.: Evolution Géodynamique D'un Bassin Intraplaque Sur Décrochements: Le Moyen Atlas (Maroc) Durant Le Méso-Cénozoïque. Travaux De l'Institut Scientifique, Rabat, no. 18, p. 142 (1989)
5. Drever, J.F.: The Geochemistry of Natural Waters, 3rd ed., p. 379. Prentice-Hall Inc., New York (1997)
6. Harmand, C., Moukadiri, A.: Synchronisme Entre Tectonique Compressive Et Volcanisme Alcalin: exemple de la province quaternaire du Moyen-Atlas (Maroc). Bull. Figure 2 Soc. Géol. France, II, 4, pp. 595–603 (1986)
7. Farki, K., Zahour, G.: Contribution to the understanding of the sedimentary and tectono-volcanological evolution of Oued Mellah (Coast Meseta, Morocco). In: Colloque International Conference of SIG Users, Taza GIS-Days, pp. 568–572 (2012)
8. Agence du Bassin Hydraulique du Bouregreg et de la Chaouia, Monographie des ressources en eau commune urbianeBniyakhlaf (2011)
9. Hamid, W.: Contribution A L'étude Tectono-Volcanique Et Sédimentologique Du Bassin Permien Et Triasique De l'Oued N'Fifikh (Meseta nord occidentale, Maroc). Mémoire de DESA, Université Hassan II Mohammedia, Casablanca (FSBM), p. 77 (2003)
10. Farki, K.: Minéralisations Ferrifères Associées Aux Formations Triasiques De l'Oued N'Fifikh (Meseta occidentale). Mémoire de Master, Université Hassan II Mohammedia, Casablanca (FSBM), p. 64 (2009)
11. Farki, K., Zahour, G., Baroudi, Z., Alikouss, S., Zerhouni, Y., El Hadi, H., Darhnani M.: Mines Et Carrières Triasico-Liasiques De La Région De Mohammedia: Inventaire, Valorisation Et Etude D'impact Environnemt, pp. 306–326 (2016)
12. Zahour, G., Farki, K., Belkhattab, H.: Carrières De Mohammedia: Impact Environnemental et Perspectives De Réhabilitation. In: 4ème édition du colloque international: argiles et environnement (Oujda, Novembre 26–28, 2010), pp. 7–8 (2010)
13. Kouassy Kaledje, P.S.: Etudes Piézométriques Et Hydrochimiques Du Bassin Versant De La Bibkola (Nord-Ouest N'Gaoundéré). DEA, Facultés des sciences, Univ. Yaoundé. Cameroun, pp. 16–26 (2010)
14. Castany, G.: Principes et méthodes de l'hydrogéologie. Bordas, Paris, p. 238 (1982)
15. Rodier, J.: Analyse Des Eaux Naturelles, Eaux Résiduaires, Eaux De Mer. 8iéme édition Du nod, Paris, p. 1384 (1996)
16. Zerouali, A.: Elaboration de la vulnérabilité moyennant la méthode DRASTIC et le Système d'Information Géographique (SIG) (cas de la nappe de Souss-Chtouka), Guide book on Mapping Groundwater Vulnerability, vol. 16 Ed. de l'IAH, p. 131 (1994)
17. Norme AFNOR: Qualité de l'eau - Échantillonnage pour analyse microbiologique, Paris (2016). https://www.afnor.org
18. Smida, H., Abdellaoui, C., Zairi, M., Ben Dhia, H.: Cartographie Des Zones Vulnérables A La Pollution Agricole Par La Méthode DRASTIC Couplée A Un Système D'information Géographique (SIG): Cas De La Nappe Phréatique De Chaffar (sud de Sfax, Tunisie). Science et changements planétaires/Sécheresse volume 21, numéro 2, avril mai et juin (2010)

19. Amrani, S., Hinaje, S.: Utilisation Des Analyses Hydro-Géochimiques Et Des Analyses en Composantes Principales (A.C.P) Dans L'explication Du Chimisme Des Eaux Souterraines De La Nappe Plio-Quaternaire Entre Timahdite Et Almis Guigou (Moyen Atlas, Maroc). Science Lib Editions Mersenne, vol. 6, no. 140306, p. 14 (2014). ISSN 2111-4706
20. Hinaje, S.: Tectonique Cassante Et Paléochamps De Contraintes Dans Le Moyen Atlas et le Haut atlas central (Midelt-Errachidia) De Puis Le Trias Jusqu'à l'actuel, p. 363. Thèse Doc. Etat, Univ. Rabat (2004)
21. Fetter, C.W.: Contaminant Hydrogeology, p. 458. Macmillam Publishing Co., New York (1993)

Towards SDI Services for Geological Maps Data

Tarik Chafiq[1(✉)], Mohammed Ouadoud[2], Hassane Jarar Oulidi[3], Ahmed Fekri[1], and Abdenbi Elaloui[1]

[1] Faculty of Sciences Ben M'Sik, University Hassan II, Casablanca, Morocco
Tarik.chafiql@gmail.com
[2] Faculty of Sciences, Abdelmalek Essaadi University, Tetouan, Morocco
[3] Hassania School of Public Works, Casablanca, Morocco

Abstract. The aim of this work is to facilitate access to geological data coming from distributed data sources and the information (metadata) related to these data. For that purpose, a spatial data infrastructure (SDI) prototype has been established which comprises a geoportal that provides access to a geological Catalogue through Catalogue Service for the Web (CSW) in order to get the metadata that describes the data as well as the available services as the WMS (Web Map Service) view and Web Feature Service (WFS) download services, which aims to be compliant with the Open Geospatial Consortium (OGC) standards rules. Indeed, the establishment of spatial data infrastructure that has been elaborated in accordance with internationally recognized standards allows the exploration and sharing of geological data information, and will also allow researchers and professionals to spend more time in the analysis and the discovery of these data. This work can be used as a model for other fields in geoscience such as geophysics and hydrology or other close fields.

Keywords: OGC standards · SDI · Metadata · Interoperability · Geology · Morocco

1 Introduction

Sustainable development requires access to data, information, knowledge and natural resources. Generally, the information have huge values for organisms regarding the profitability of these data for their own background [1]. Indeed, this type of data referred to as geographical or spatial data [2], which represents an essential element of the knowledge available to the modern science and technology of information and communication, helps to take a decision in both the public and private spheres.

In the last decade, many organizations and Moroccan institutes participated in the gathering of Geoscience information for different purposes. This information is vital for resource management in various fields (e.g., natural resources, land registers, business, etc.), and they provide answers for specific scientific questions or help to test a certain hypothesis [3] at different levels (i.e., local, regional, national and global), and contributes to better environmental knowledge. Unfortunately, most of this data cannot be used due to the lack of details related to the information (metadata), which is a key in

© Springer Nature Switzerland AG 2019
M. Ezziyyani (Ed.): AI2SD 2018, AISC 913, pp. 34–44, 2019.
https://doi.org/10.1007/978-3-030-11881-5_4

supporting the discovery, evaluation and application of geographic data beyond the original organization or a specific project, and different standards adopted by each organization. It is commonly known that approximately 50% of time is lost in searching data while doing environmental assessments [4]. Therefore, facilitating data discovery across the various parties interested in geospatial information will certainly help to lower this percentage and will give more time to data analysis, which is a crucial step to better understand complex environmental issues and interactions.

The fundamental requirement for an efficient and effective data discovery mechanism is that data should be properly documented with metadata and stored in a catalog [5]. Otherwise, without appropriate metadata an SDI will fail in its main objectives; facilitating geospatial data discovery and the access to it [6].

In this study, as a response to the listed needs, some Geospatial technologies were used within the framework of an SDI in Morocco. They represent a rapid and cost-effective tool in producing valuable data on geosciences [7]. These technologies aim to solve the aforementioned problems, in particular by promoting and enabling the sharing and access to geological maps and metadata among two data providers or more, and thus, contributes to the harmonization of Geo-information. This approach reverses a trend for the dissemination of knowledge and transfer of geological information using open source technologies which can be achieved from a pure technical perspective and apart from organizational arrangements, appropriate policies of the core components of SDIs.

2   Related Works

Recently, the management and sharing of geological information using GIS techniques on the Web have been quickly increased and developed. So, they can more effectively manage geological information.

The OneGeology-Europe project gathers 29 partners from 20 nations [8]. It aims to make the first geological maps at a scale 1: 1000000 (1:1 M) of Europe discoverable and accessible through a single geoportal. It provides access to data and metadata held by each organization, and is based on standards developed in the Onegeologie-Europe project and under a common license. The OneGeology-Europe portal management is now the responsibility of the EGS (European Geological Surveys). It contacted the Bureau de Recherches Géologiques et Minières (BRGM) to maintain the GeoPortal, and Czech Geological Survey (CGS) to maintain the catalogue of metadata [9].

The experience of the unity of Italy Torino IGG (Institute of Geosciences and Earth Resources) is an example of a Spatial Data Infrastructure that allows the transfer of knowledge of local research activities to the web [10]. The Geoportal of IGG was built as part of the IDE-Univers project [11], and then implemented under the initiative of the CNR (National Research Council) [12]. Indeed, the specific objective of IDE-Univers project support access, sharing and interoperability of the geological information produced by universities and research centres. Consequently, it leads the data providers to bring out geological interpretations as much as possible in explicit format predefined in Catalog Manager.

3   Switching to Digital Mapping

An important transition took place there about 25 years, when traditional geological maps have gradually switched to digital format (digital-format) [13]. The rapid development of digital technology has significantly improved how to create, manage and share geological maps with the use of geographic information systems (GIS). This represents a considerable analysis and potential valuation that transformed geological maps from traditional static documents to dynamic form in databases. In this case, information is systematically recorded, operated and can be represented differently depending on the needs. In addition, it allows the analysis, downloading and sharing together with other distributed data sources. GIS has first been carried out as a desktop tool, their meanings have also increased by the development of a related field, for instance, web GIS and mobile GIS.

4   OGC Standards and Geological Information Sharing

The geological data of many digital mapping projects should be standardized, archived, structured and properly used through an adequate and effective system for the management of applications [14]. This should be particularly used in case of natural resources management. Applications using Web-based GIS are essential in order to maximize the sharing of geological information and to help geologists and non-geologists retrieve geological information [15]. In fact, the webcast of the geological maps strategy is always supported by information technological. The maps are published in the form of digital images on dedicated Websites through web services and interactively retrieved via the geoportals. These can be considered as a gateway to geographical and geological resources. It is the Web environment that allows a community of users to discover and share content and Geospatial services [16]. Generally, it can be integrated in spatial data infrastructure (SDI) [17, 18], i.e. the OneGeology-Europe initiative [9].

Digital geological maps are shared and exploited via geoportals form images through web services such as WMS (Web Map Service). This is compatible with OGC (Open Geospatial Consortium) that defines the WMS as a simple HTTP interface. It makes enable the production of dynamic cartographic representations as images (JPEG, PNG, etc.) scanned and georeferenced from distributed geospatial databases, and can be displayed in a browser application [19] or/and desktop [19]. The design type WFS (Web Feature Service) allows to transmit vectorial data via the Internet to GML (Geographic Markup Language) [20] and to make transactions via the data network infrastructure [21]. In fact, the web services listed above are independent of programming languages, operating systems, etc. Hence, the technological interoperability is effectively ensured.

In geology, two factors are of fundamental importance. First, the representation of the geological data. Second, the availability of its documentation (metadata).

- Simon Cox, researcher at the CSIRO Exploration and mining Division and Nick Ardlie, Geoscience Australia, have both been involved in the development of

GeoSciML [22]. GeoSciML is a specific geoscience data model based on Geography Markup Language (GML) for the representation of the features and geometry [23]. It provides the storage and exchange of Geoscience information [24, 25], and aims to enable the harmonization of the geological data according to different types of database. Moreover, it describes the geological information, which must be made available through the web services consistent with the infrastructure for the exchange of geological data on the Internet.

- Geological data documentation efforts were supported by the use of standards, considered important keys to facilitate the exchange of metadata, interpreted by individuals and manipulated by machines. Metadata usually play a key role in resource description and discovery [26], and represent the information on data [27, 28]. They help to create an environment to meet today's demand [29]. In fact, there is a web OGC cataloguing specific service that interacts with one or more catalogues of spatial resources, usually via distant CSW [30]. It describes the specification of web service interface for the discovery of standardized metadata and provides information that describes the data, production processes, conditions of use and related information such as the accuracy and content description. The definition of the set of metadata elements is mandatory according to ISO 19139 (2007) [32]. Geospatial metadata aims to define the set of specific geographical data. As noted in SDI cookbook [31]: 'metadata helps people who use geological data to find the information they need to determine how best to exploit them'. They can document the essential characteristics of services or Geospatial applications, where it responds to the 'what, why, when, who, where and how' questions about the resource. These issues include [31]:
  - What: Title and description of information;
  - Why: A summary describing the reasons for the collection of data and its uses;
  - When: When the data were created and the update cycles;
  - Who: originator, data supplier, and possibly intended audience;
  - Where: The geographical extent outcome of latitude and longitude coordinates, geographical names or administrative areas;
  - How: How the data was generated and how to access the data.

The use of common standards attenuates many difficulties that can be encountered when sharing data [32]. In fact, they allow geological data to be better processed, preserved, recombined and reused in different contexts via a geoportal.

5 Moroccan Prototype

To facilitate the discovery and access to geoscience information, SDIs provide the framework for data publishing, discovery, gathering and integration, which enable interoperability of the different components involved at different organization levels (e.g., regional, national, or global level) and involving both public and private institutions [33]. Hence, the establishment of a spatial data infrastructure for the Moroccan prototype for the exploitation of geological data is entirely based on open source technologies with community standards of the OGC and its services to ensure the interoperability. Indeed, this solution is a part of IT infrastructure consisting of a set of servers including a map server, two servers for database management, and a metadata management system.

Geological maps that are the subject of our application test represent a sample of two geological maps at 1:50,000 that are neighboring Qal'at Mgouna [34] and Sidi Flah [35]. They cover partially a set of 6 geological maps produced in collaboration with the USGS in the National Plan for Geological Mapping.

5.1   Qal'at Mgouna

The three quarters of the surface of the territory covered by the geological map of Qal'at Mgouna at 1:50,000 is located in the northern flank of Central Saghro, and on the plains formed by the valley of Oued MGOUN and Oued DADES. The coordinates of this area are (31°00 N and 31°15 N) latitude and (6°00 W and 6°15 W) longitude (see Fig. 1).

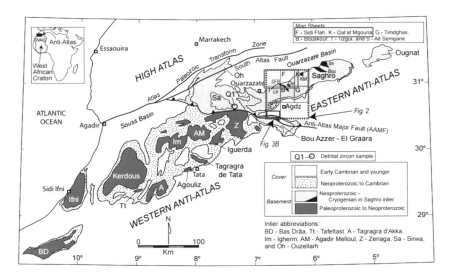

Fig. 1. Geological map of Western Anti-Atlas [36]

5.2   Sidi Flah

The third of the surface of the territory covered by the geological map of Sidi Flah at 1:50,000 is located in the northern flank of Central Saghro, and on the plains formed by the valley of Oued DADES which flows from the NE to the SW and the valleys of Oued Imassine, Imarighene and Issil-n-Ibghell. The coordinates of this area are (31° 00 N and 31°15 N) latitude and (6°15 W and 6°30 W) longitude (Fig. 1).

6 Methodology

In this section, we describe the methodology that has been used for exploration and exploitation of geospatial information through web services implemented in Moroccan prototype.

6.1 Technical Management Information

The cards aforementioned are from two different databases of servers represented by Oracle and PostgreSQL, that uses spatial extensions (PostGIS, Oracle Spatial) to manage the types of geometric and geographic data [37]. These two databases are directly linked to the map server to publish this information online through a client side interface (Web-Mobile-Desktop) that allows visualising, researching and exploring geospatial data.

Making geological and geographical information available, involves the use of a metadata servers or simply a Geocatalogue [38]. The need to disseminate spatial information on the Internet has led to the creation of certain solutions dedicated to this goal [39]. In addition, the flexibility of the Internet has helped develop more powerful solutions namely PyCSW, GeoNetwork, Geosource and others to register, download and upload of data related to spatial data. In this study GeoNetwork has been chosen as a catalogue for our prototype.

GeoNetwork is an Open Source solution which was used as a central repository for ISO 19115/19119 [40], and also designed to build catalogues of spatial data that respond to general questions posed above (what? Why? When?...etc.). Thus, metadata in the catalogue corresponds to the OGC standards that provide interoperability with other services of catalogues on the Internet.

The GeoNetwork in our prototype allowed us as administrator to create metadata (see Fig. 2) and also gave the possibility to users to search (see Fig. 3) according to their needs in addition to their contributions. In fact, users can pass through the search engine that returns all words that contain the character string which is querying, which is called IntelliSense, or through the filter of registry in order to have the information in different formats namely, EXtensible Markup Language (XML), Keyhole Markup Language (KML) or others.

The catalogue aims to ensure continuity and interoperability to other catalogues in order to manage sources for spatial references between the spatial information communities and their data [41]. Indeed, these catalogues services are vital and of highest priority for the success of SDI list descriptions about geospatial resources, namely the metadata records.

The map server in our prototype plays a very important role in the manipulation of spatial data. It allows sharing analysis and publishing of geospatial from a variety of sources. It is managed by the GeoServer, which is one of the biggest players in the field of web mapping [42].

Fig. 2. The metadata editor of Moroccan prototype

Fig. 3. Searching for available resources

6.2   Architecture and Visualization

For the correct visualization of geospatial data (geological maps, topographic maps…), the GeoExplorer has been used as a database viewer application, included with the distribution OpenGeoSuite Community Edition, and based on OpenLayers and GeoExt which does not depend on the components of proprietary software. Moreover, it focuses on the display of the maps, the management of layers either WMS or WFS, so it enables users to have information in the form of an attribute related to geospatial data table. In fact, it gathers the maps GeoServer and integrates the funds on the basemap hosted, namely Google Maps or OpenStreetMap. Communications between GeoServer and GeoExplorer are obtained via WFS and WMS protocols.

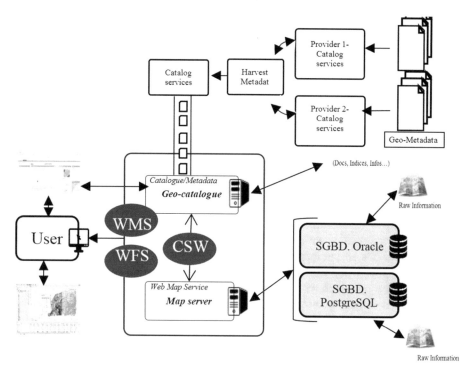

Fig. 4. Architecture of Moroccan prototype

When the tools and architecture are combined as mentioned earlier (see Fig. 4), the geospatial data will be easily accessible and exploitable by the different kinds of users mainly organizations, institutions…etc., through geoportal of Moroccan prototype.

7   Discussion

Even if geoscience information can bring major benefits for the economy and development of developing countries alike, most of them are still lacking timely access to proper geospatial data [43] if it does already exist or accessible. Thereby, some others [44] emphasize the necessity of having tools that facilitate the management of digital and the publication of this metadata on the Internet. For example, the rate of the exchange of geoscience information is still very weak in Morocco, knowing that it has introduced The article 27 which reads "every citizen has the right to public information" [45] in its newly adopted constitution in 2011. And in 2015, the final draft Law No. 31-13 that guaranties the exercise of article 27 was submitted to the parliament [46].

The findings of this research will facilitate the sharing of geological and geographical data with their metadata, which will enhance the Maroccan geographic information catalog that is compatible with the profile of the OpenGIS specification. The sharing of and access to this information was almost impossible in spite of the existence of this information in cookbooks entitled "notes and memoirs of the geological survey of Qal'at Mgouna" [34] and "notes and memoirs of the geological survey of Sidi Flah" [35]; because of the lack of tools that will allow access to them. The Moroccan SDI prototype that has been established will enable us to overcome the aforementioned problem of sharing and accessing information. Therefore, the Moroccan SDI prototype that in some way drives data makers to bring out geological interpretations as much as possible in explicit format.

8 Conclusion

The presented approach to Moroccan prototype integrating Open Source Software for implementing an SDI is successful. As outlined before, the integration of several Open Source technologies, which are market leaders and reference implementations in their own domains, is a recommendable approach with many advantages. In fact, this approach allowed us to share geological and geographical data with their information (metadata) and related interpretations which are usually not entirely stored in a database. In fact, this work has been considered within this framework not only as a dynamic application which will grow with the contributions of the users, but also as an effective option to meet the needs of public and private organizations that operate in the field of geosciences.

Further analysis will be carried out on metadata harvested from the OGC Service discovered in order to give a more accurate preview of the number of OGC services, which are not recorded in the CSWs available in the next step. Another extension of this work will be conducted through the promotion of principles of data related to the normal connection that is semantically related to geospatial information resources implemented.

References

1. Chafiq, T., Groza, O., Oulid, H.J., Fekri, A., Alexandru, R., et al.: Spatial data infrastructure. Benefits and strategy. Analele stiintifice ale Universitatii "Alexandru Ioan Cuza" din Iasi-seria Geografie **61**, 21–30 (2015)
2. Ajmi, M., Hamza, M., Labiadh, M., Yermani, M., Khatra, N.B., et al.: Setting up a spatial data infrastructure (SDI) for the ROSELT/OSS network. J. Geogr. Inf. Syst. **6**, 150 (2014)
3. Ludäscher, B., Lin, K., Bowers, S., Jaeger-Frank, E., Brodaric, B., et al.: Managing scientific data: from data integration to scientific workflows. Geol. Soc. Am. Spec. Pap. **397**, 109–129 (2006)
4. Craglia, M., Campagna, M.: Advanced regional SDIs in Europe: comparative cost-benefit evaluation and impact assessment perspectives. Int. J. Spat. Data Infrastruct. Res. **5**, 145–167 (2010)
5. Charvat, K., Vohnout, P., Sredl, M., Kafka, S., Mildorf, T., et al.: Enabling efficient discovery of and access to spatial data services. Int. J. Adv. Comput. Sci. Appl. **3**, 28–31 (2013)
6. Masser, I.: The future of spatial data infrastructures, pp. 14–16 (2005)

7. Mayilvaganan, M.K., Mohana, P., Naidu, K.: Delineating groundwater potential zones in Thurinjapuram watershed using geospatial techniques. Indian J. Sci. Technol. **4**, 1470–1476 (2011)
8. Onegeology-Europe Partners. Onegeology-Europe
9. Čápová, D.: OneGeology-Europe Plus Initiative – Final report (2014)
10. Balestro, G., Sara, B., Fabrizio, P., Sergio, T.: Pubblicazione e condivisione di banche dati geologico-strutturali mediante tecnologie Web Map Service. Rendiconti Online **8**, 12–13 (2009)
11. Poggioli, D., Sara, R., Carrara, P., Bertozzi, R., Montaguti, M., Barea, M., Guimet, J., Redondo, M., vaitis, M.N.: IDE-Univers: una infrastruttura di dati spaziali tematica per Università e Centri di Ricerca. Atti 11a Conf. Naz. ASITA (2007)
12. NR Institute of geosciences and earth resources
13. Whitmeyer, S.J., Nicoletti, J., De Paor, D.G.: The digital revolution in geologic mapping. GSA Today **20**, 4–10 (2010)
14. Chang, Y.-S., Park, H.-D.: Development of a web-based geographic information system for the management of borehole and geological data. Comput. Geosci. **30**, 887–897 (2004)
15. Ma, X.: Ontology spectrum for geological data interoperability (2011)
16. Maguire, D.J., Longley, P.A.: The emergence of geoportals and their role in spatial data infrastructures. Comput. Environ. Urban Syst. **29**, 3–14 (2005)
17. Hennig, S., Belgui, M.: User-centric SDI: addressing users requirements in third-generation SDI. The Example of Nature-SDIplus. Geoforum Perspektiv **10**, 30–42 (2013)
18. Iwanaik, A., Kaczmarek, I., Kubik, T., Lukowicz, J., Paluszynski, W., et al.: An intelligent geoportal for spatial planning (2011)
19. Michaelis, C.D., Ames, D.P.: Considerations for implementing OGC WMS and WFS specifications in a desktop GIS. J. Geogr. Inf. Syst. **4**, 161 (2012)
20. Genot, V., Buffet, D., Legrain, X., Goffaux, M.-J., Cugnon, T., et al.: Pour un échantillonnage et un conseil agronomique raisonné, les outils d'aide à la décision. Biotechnologie Agronomie Société et Environnement **15**, 657–668 (2011)
21. Vretanos, P.A.: Web feature service implementation specification. Open Geospatial Consortium Specification 04-094 (2005)
22. Cox, S., Ardlie, N.: Geoscience Australia AND CSIRO developing the GeoSciML interoperability standard with Enterprise Architect (2009)
23. ISO 19136, Geographic information—Geography Markup Language (GML) 394 (2007)
24. Asch, K., Jackson, I.: Commission for the management & application of geoscience information (CGI). Episodes **29**, 231–233 (2006)
25. Tegtmeier, W., Zlatanova, S., van Oosterom, P.J.M., Hack, H.R.G.K.: 3D-GEM: geotechnical extension towards an integrated 3D information model for infrastructural development. Comput. Geosci. **64**, 126–135 (2014)
26. Chen, Y.-N., Seadle, M., Seadle, M.: A RDF-based approach to metadata crosswalk for semantic interoperability at the data element level. Library Hi Tech **33**, 175–194 (2015)
27. Ball, C.: Beyond data about data: The Litigator's guide to metadata (2005)
28. Hey, A.J., Trefethen, A.E.: The data deluge: an e-science perspective (2003)
29. Guptill, S.C.: Metadata and data catalogues (1999)
30. OGC: OpenGIS Catalogue Service Implementation Specification 2.0. 2 (2007)
31. 3GSDI: The SDI CookBook (2004). http://www.gsdiorg/gsdicookbookindex
32. Ball, A., Chen, S., Greenberg, J., Perez, C., Jeffery, K., et al.: Building a disciplinary metadata standards directory. Int. J. Digit. Curation **9**, 142–151 (2014)
33. Nebert, D.D.: Infrastructure GSD. The SDI cookbook: developing spatial data infrastructures. Global Spatial Data Infrastructure (2001)

34. Stone, B., Benziane, F., El Fahssi, A., Yazidi, A., Walsh, G., et al.: Carte géologique au 1/50 000, Feuille Kelâat M'gouna. Notes et Mémoires du Service Géologique du Maroc 468 (2008)
35. Stone, B., Benziane, F., El Fahssi, A., Yazidi, A., Walsh, G., et al.: Carte géologique au 1/50 000, Feuille Sidi Flah. Notes et Mémoires du Service Géologique du Maroc **467**, 114 (2008)
36. Choubert, G.: Histoire géologique du Précambrien de l'Anti-Atlas: Éditions du Service géologique du Maroc (1964)
37. Kazemitabar, S.J., Demiryurek, U., Ali, M., Akdogan, A., Shahabi, C.: Geospatial stream query processing using Microsoft SQL Server streamInsight. Proc. VLDB Endow. **3**, 1537–1540 (2010)
38. Farazi, F., Maltese, V., Giunchiglia, F., Ivanyukovich, A.: A faceted ontology for a semantic geo-catalogue. The semanic web: research and applications, pp. 169–182. Springer (2011)
39. Stefanakis, E.: Map mashups and APIs in education. In: Online Maps with APIs and WebServices, pp. 37–58. Springer (2012)
40. Kliment, T., Granell, C., Cetl, V., Kliment, M.: Publishing OGC resources discovered on the mainstream web in an SDI catalogue, pp. 14–17 (2013)
41. Poulet, T., Corbel, S., Stegherr, M.: A Geothermal Web Catalog Service for the Perth Basin (2010)
42. Youngblood, B.: GeoServer Beginner's Guide. Packt Publishing, Limited (2013)
43. Ayanlade, A., Orimoogunje, I., Borisade, P.: Geospatial data infrastructure for sustainable development in sub-Saharan countries. Int. J. Digit. Earth **1**, 247–258 (2008)
44. Giuliani, G., Papeschi, F., Mlisa, A., Lacroix, P.M.A., Santoro, M., et al.: Enabling discovery of African geospatial resources. South East. Eur. J. Issue Earth Obs. Geomat. **4**, 1–16 (2015)
45. UNESCO: UNESCO supports freedom of information in Morocco. UNESCO (2013)
46. Ayoub, L.: Droit d'accès à l'information. Le Matin (2016)

Modeling and Optimization of Operating Parameters for Removal of Acid Red35 Dye from Reconstituted Waste Water by Electrocoagulation in an Internal Loop Airlift Reactor Using the Experimental Design

N. Azeroual[(⊠)], A. Dani, B. Bejjany, H. Mellouk, and K. Digua

Laboratory of Process Engineering and Environment,
Faculty of Science and Technology of Mohammedia,
Hassan II University of Casablanca, Casablanca, Morocco
naimaazeroual@yahoo.fr

Abstract. In this study, we investigated the optimization of the treatment, in an internal loop airlift reactor (ILAL Reactor), by electrocoagulation (EC) to remove AcidRed35 (AR35) Dye from aqueous solutions using the Aluminum anode, and taking into account many factors such as electric conductivity, applied voltage, the treatment time and the inter-electrode distance Dint.

To verify these factors and their effects on the (EC) of (AR35), we have established a model following the Methodology of Experimental Design. The mathematical model is established, initially using a full screening plan, to verify the existence of the effect of these four factors, and in a second time, a central composite design (CCD) is applied [1]. The model describes the change in measured responses, of dye removal efficiency (R (%)), and energy consumption (E cons (W h/m^3)), according to the three factors (the conductivity, the treatment time and the inter-electrode distance (Dint)). The voltage was fixed at 9 V because, it have no effect on the removal efficiency of (AR35) during the screening step.

The graphical representation of this model, in the variable space, and "desirability function" allowed us to define the optimal conditions for these parameters. The optimum value of the conductivity, the time, and Dint are respectively 1000 µs/m^2, 20 min and 1.5 cm. A 53% efficiency of dye removal is observed with an energy consumption of 1859 Wh/m^3, and this optimal condition is confirmed experimentally for the two responses.

Keywords: The Acid Red35 dye · Electrocoagulation · Aluminum anode · Internal loop airlift reactor · Methodology of plans of experiments

1   Introduction

Polluted water from finishing, dyeing and washing lines in the industry of textile is a big source of pollution. This liquid effluent usually has an intense color, because most dyes are visible in water, even at very low concentrations (<1 mg.L^{-1}) [2]. Thus, they

© Springer Nature Switzerland AG 2019
M. Ezziyyani (Ed.): AI2SD 2018, AISC 913, pp. 45–60, 2019.
https://doi.org/10.1007/978-3-030-11881-5_5

contribute to the pollution problems related to the generation of a considerable quantity of wastewater, containing residual dyes [3]. This is largely due to the degree of fixation of theses dyes to cellulosic fibbers. The discharge of these wastewaters into the ecosystem is a dramatic source of pollution, and non-aesthetic disturbance in aquatic life; therefore, it presents a potential bioaccumulation hazard that can affect human health. These waters must therefore, be treated before their discharge, for environmental and legal reasons.

This subject is particularly critical in Morocco, as the industry of textile is in full expanding. Conventional methods, generally used for the removal of dyes from industrial polluted waters, are mainly biological and physicochemical treatments and their various combinations [4, 5]. Nevertheless, biological treatments are the most used.

The toxicity of dyes usually poses the problem of bacterial growth, thus limiting the effectiveness of the discoloration [5]. Physicochemical methods are generally based on adsorption, coagulation/flocculation, chemical oxidation and photo degradation (UV/H_2O_2, UV). However, these treatment methods usually consume a large amount of chemicals, which sometimes lead to secondary pollution and a great volume of recovered solids (sludge) [6]. It has recently been shown that the techniques of treating polluted water, based on the non-polluting aspect and the automation facilities that electricity brings, are economically attractive, avoid the majority of these problems and should allow significant development of electrochemical techniques for the treatment of these dyes. Many techniques using directly or indirectly, electrical energy have already developed [1]. Among them, Electrocoagulation has been used successfully to treat a variety of industrial wastewater [7–11]. The purpose of this method is to form flocs of metal hydroxides in the effluent to be treated by electro-dissolution of soluble anodes. The main reactions to Aluminum electrodes are:

$$\text{Anode}: Al \rightarrow Al^{3+} + 3e-$$
$$\text{Cathode}: 2H_2O + 2e- \rightarrow H2 + 2OH-$$

Al_{aq}^{3+} and OH– generated by the electrode reactions above, react to form different monomer and polymer species which finally transform into $Al(OH)_3$ [1].

Electrocoagulation has the potential to be the distinct economic and environmental choice for wastewater treatment [1]. It requires simple equipment designed for virtually any size. Start-up and operating costs are relatively low [9]. It produces minimal mud.

The field of application of the airlift is very wide especially in industrial processes that require a large transfer of material and a good mixture, for example: in the field of biological treatment of wastewater and industrial effluents, and in the chemical industry and biotechnology [12].

In this study, the treatment of the AR35 textile dye by EC is carried out in an ILAL reactor with a reaction volume of 850 mL. This type of reactor has been chosen as enclosure for EC of the target compound to avoid energy supplement do to the stirring that is assured in the airlift only by generation of the bubbles of hydrogen. The effect of electric conductivity, applied voltage, treatment time and distance between electrodes on EC efficiency was studied by following the evolution of the color removal efficiency and the energy consumption by drawing up the plan of experiences [1–14].

2   Materials and Methods

2.1   Reagents

Acid Red35 dye: The Acid Red35 dye (CAS: 6441-93-6), is in the form of a red powder of chemical formula $C_{19}H_{15}N_3Na_2O_8S_2$. Its chemical structure is given in Fig. 1, and molar mass is 523.45 g/mol. It is used as a target molecule to simulate wastewater from the textile industry. The solutions that are the subject of this study are prepared by dissolving 20 mg of the AR35 dye in one liter of distilled water so as to obtain an initial concentration $C_0 = 20$ mg.L^{-1}.

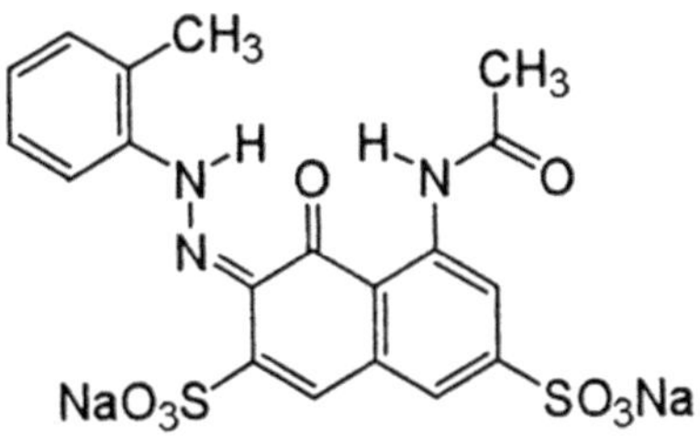

Fig. 1. Chemical structure of AR35 dye

2.2   Methods of Analyses

The reduction of coloration was followed by UV/visible spectrophotometry. The spectrophotometer used is brand of HITACHI, model U5100 using a quartz cuvette with a width of 10 mm.

The measurement of the absorbance at 515 nm for dye concentrations of between 0 and 20 mg.L^{-1} made it possible to obtain a calibration curve whose evolution is linear. Therefore, the residual dye concentration is deduced from (1).

$$C = 63,\ 03\ A \tag{1}$$

The coefficient of determination R^2, for the establishment of this linear regression equation, is almost equal to one ($R^2 = 0.9996$). Therefore, using this equation can correctly determine the concentration by measuring the absorbance with a very small error.

The performance of the AR35 dye treatment is evaluated by the calculated dye removal efficiency using (2).

$$R(\%) = (C_t - C_0)/C_0 * 100 = (A_t - A_0)/A_0 * 100 \tag{2}$$

With, C_0 and A_0 are, respectively, the dye concentration and the absorbance at the wavelength 515 nm at ($t = 0$ s).

C_t and A_t are respectively the dye concentration and the absorbance at the wavelength 515 nm at an instant t corresponding to a treatment time t.

The energy consumption $(E_{cons}(Wh/m^3))$ in the EC process results mainly in Ohmic losses rather than electrolysis reactions, and calculated from (3):

$$E_{cons}(Wh/m^3) \ = \ U \times I/(Q \times 1000) \tag{3}$$

Where U is the voltage (V), I is the electrolysis current (A) and Q is the volume flow (m^3/h).

To design, model and optimize the experimental design, we used the JMP Software (SAS JMP, Statistical Discovery 11.0).

3   Experimental Apparatus

The Electrocoagulation tests were conducted in a rectangular ILAL reactor shown schematically in Fig. 2. The dimensions of the reactor are as follows: Reaction volume: 850 mL, useful height (a) 36 cm, width (b) 5.5 cm, height injection area (c) 4 cm, height recirculation zone (d) 2.3 cm. Flat Aluminum bars are used as electrodes, with a total height of 50 cm, a thickness of 0.2 cm and a width of 2.5 cm. The surface of the anode is 90 cm^2. These bars are totally immersed in the Riser, and subjected to a voltage thanks to a DC voltage generator.

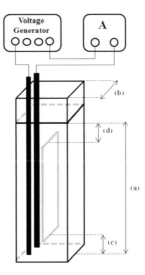

Fig. 2. Schematic of the laboratory scale airlift reactor system

4   Results and Discussions

The ability of EC to remove (AR35) Dye from aqueous solutions is investigated in an ILAL reactor using the Aluminum anode and taking into account many factors such as conductivity, voltage, time and inter-electrode distance. To verify these factors and their effects on the electrocoagulation of (AR35), we established a model following the methodology of the experimental design.

4.1 Full Factorial Plan

The chosen experimental design is full factorial plan with four factors and two levels. The main objective is to verify the existence of the effect of the four independent variables: conductivity, potential, time and (Dint).

Experimental Domains of Factors. The experimental domains were determined following preliminary tests. Table 1 shows the experimental values of the four independent variables X_1 (conductivity) X_2 (the applied voltage) X_3 (time) and X_4 (D int).

Table 1. Experimental domains of the factors studied for Full Factorial Plan

Variables	Factors	Experimental domain	
		−1	+1
X_1	Electrical conductivity (µS/cm)	1000	2000
X_2	Applied voltage (V)	6	12
X_3	Time (min)	13	31
X_4	D int (cm)	1.4	2.2

The experimental plan was then designed from 16 experiments.

Responses Studied. For this plan, the responses studied are the AR35 dye removal efficiency (Y_1), and the energy consumption (Y_2).

Experiments Matrix. The matrix of experiments of the factorial plan 2^4 is listed in Table 2.

Table 2. Experimental matrix of full factorial design 2^4

N Exp	N Rand	Configuration	X_1	X_2	X_3	X_4	Y_1 (%)	Y_2 (Wh/m^3)
1	11	----	1000	6	13	1.4	42.4	436.43
2	7	+---	2000	6	13	1.4	33.33	776.22
3	2	-+--	1000	12	13	1.4	46.31	1929.08
4	5	++--	2000	12	13	1.4	48.53	3347.95
5	10	--+-	1000	6	31	1.4	47.43	1045.09
6	13	+-+-	2000	6	31	1.4	50.3	1863.12
7	15	-++-	1000	12	31	1.4	63.13	4600.61
8	14	+++-	2000	12	31	1.4	68.24	7936.82
9	9	---+	1000	6	13	2.2	15.74	251.04
10	1	+--+	2000	6	13	2.2	37.92	536.8
11	16	-+-+	1000	12	13	2.2	11.84	704.94
12	4	++-+	2000	12	13	2.2	29.59	1589.32
13	6	--++	1000	6	31	2.2	40.74	607.54
14	3	+-++	2000	6	31	2.2	47.4	1291.84
15	8	-+++	1000	12	31	2.2	33.88	1844.77
16	12	++++	2000	12	31	2.2	39.05	3988.85

Mathematical Model. The mathematical model representing each response is a first-order polynomial given by the following expression:

$$Y = b_0 + b_1x_1 + b_2x_2 + b_3x_3 + b_4x_4 + b_{12}x_1x_2 + b_{13}x_1x_3 + b_{23}x_2x_3 + b_{14}x_1x_4 + b_{24}x_2x_4 + b_{34}x_3x_4 \quad (4)$$

Results of the Experimental Plan

Statistical Validation of Models. Table 3 represents the ANOVA table obtained for the two responses. The coefficients of determination for Y_1 and Y_2 responses were respectively 95 and 98%. Both models (Y_1 and Y_2) showed good agreement between experimental and predicted values. This implies that 95 and 98% of the sample variation for dye removal efficiency and energy consumption, respectively, are explained by the model.

Table 3. ANOVA table of Y_1 and Y_2 responses

Source of variation	Sum of squares	Ddl	Middle square	F report	p-Value
Y_1					
Regression	3059.5345	10	305.953	9.0352	**0.0127***
Residues	169.3112	5	33.862		
Total	3228.8457	15			
R^2	95%				
Y_2					
Regression	60985878	10	6098588	26.5475	**0.001***
Residues	1148617	5	229723		
Total	62134495	15			
R^2	98%				

* Statistically significant at a level of confidence greater than 95% ($p < 0.05$).

ANOVA analysis showed that both p-values are less than 0.05 (0.01 and 0.001) indicating that the two final models are significant at a 95% confidence level.

Mathematical Models of the Two Responses. The estimation of the coefficients of the first order polynomial model has been made on the basis of the least squares method.

The Fisher's 'F' test is used to determine the significance of each interaction among the variables, which in turns may indicate the patterns of the interactions among the variables.

In general, the larger the magnitude of F, the smaller the value of P, the more significant is the corresponding coefficient term [1–14].

Table 4 shows the estimation of the coefficients in the model expressions.

Table 4. Estimation of model coefficients that relate responses to factors

Term	Y_1			Y_2		
	Estimation	Rapport F	Prob. > F	Estimation	Rapport F	Prob. > F
b_0	40.989375		**<0.0001***	40.989375		**<0.0001***
b_1	3.305625	5.1631	*0.0722*	3.305625	5.1631	**0.0036***
b_2	1.581875	1.1824	0.3265	1.581875	1.1824	**0.0002***
b_3	7.781875	28.6136	**0.0031***	7.781875	28.6136	**0.0009***
b_4	−8.969375	38.0127	**0.0016***	−8.969375	38.0127	**0.0021***
b_{12}	0.475625	0.1069	0.7570	0.475625	0.1069	**0.0319***
b_{13}	−0.829375	0.3250	0.5933	−0.829375	0.3250	0.0881
b_{23}	0.721875	0.2462	0.6408	0.721875	0.2462	**0.0087***
b_{14}	3.164375	4.7313	0.0816	3.164375	4.7313	0.3639
b_{24}	−5.011875	11.8687	**0.0183***	−5.011875	11.8687	**0.0077***
b_{34}	0.465625	0.1024	0.7619	−269.0663	5.0424	0.0747

* Statistically significant at a level of confidence greater than 95% ($p < 0.05$).

It is observed that the linear effect coefficients of the time and Dint for the dye removal efficiency are very significant (Prob < 0.05). The conductivity (Prob = 0.0722) is less significant and the potential is not significant (Prob = 0.3265).

No interaction effect of the variables, except the interaction between the potential and D_{int} (Prob = 0.0183) is significant.

However, for energy consumption, all linear effect coefficients, conductivity, potential, time, and D_{int} (Prob < 0.05) are very significant.

The interaction effect of the variables is considered very significant, except the interaction between conductivity and time (P = 0.0881) and the interaction between the conductivity and D_{int} (P = 0.3639).

The quadratic effect of all variables is not observed for both responses. Therefore, the best-fit response functions for both models are written as follows:

$$Y_1 = 40.989 + 3.30X_1 + 7.78X_3 - 8.969X_4 - 5.0118\ X_2X_4 \tag{5}$$

$$Y_2 = 2046.901 + 619.463X_1 + 1195.891X_2 + 850.428X_3 - 695.013X_4 + 353.478X_1X_2 \\ + 499.541X_2X_3 - 515.808X_2X_4 \tag{6}$$

As shown in Fig. 3, the dye removal efficiency increases with increasing time and decreasing D_{int}. For conductivity, there is little effect on the removal efficiency and the potential has no effect.

The energy consumption increases with each parameter. The energy consumption is high for a higher electrolysis potential. The same trend was found by Zaroual et al. [14].

We also note that the influence of potential is not significant. Then, to have a low energy, it is advantageous to work with a low potential. The lowest potential value that gives the maximum dye removal is 9 V.

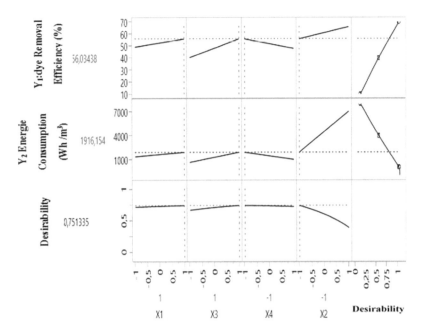

Fig. 3. Graphical representation of desirability functions associated with the responses: Y_1 and Y_2

Conclusion. The most important parameters that affect EC efficiency are conductivity, time and D_{int}. The potential does not have a significant effect on the removal efficiency and is set at 9 V.

In order to study the combined effect of these three remaining factors. Experiments are performed for different combinations of parameters using the composite central Design for three factors. Uniformed precision was selected thus 20 experiments were conducted.

4.2 Central Composite Design (CCD)

To find the optimal dye removal conditions in an internal loop airlift reactor, the experimental design should be determined based on the selected key factors. In this study, the chosen plan is the central composite design (CCD) for three factors with six repetitions in the center. The goal is to analyze the effect of the three factors, namely: conductivity, time and Dint on the two chosen responses.

As shown in Fig. 4, this rotating experimental design was performed as a (CCD) consisting of 20 experiments (uniformed precision). For three variables (n = 3), the total number of experiments was determined by the expression: 2n ($2^3 = 8$: factor points) + 2n (2 * 3 = 6: axial points) + 6 (center points: six repetitions).

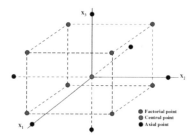

Fig. 4. Schematic diagram of the central composite design (CCD) versus X1 (Conductivity), X2 (time) and X3 (inter-electrode distance) according to factor plan 23 with six axial points and six center points (replication) (fixed factors: The potential is fixed at 9 V, total volume 850 mL).

Experimental Domains of Factors. The experimental domains were determined following preliminary tests. Therefore, the selected levels are indicated in the Table 5: Table 5 shows the experimental values of the three independent variables X_1 (Conductivity), X_2 (time) and X_3 (D_{int}).

Table 5. Experimental domains of the factors studied for (CCD)

Variables	Factors	Experimental domain				
		a	−1	0	+1	A
X_1	Electrical conductivity (µS/cm)	660	1000	1500	2000	2340
X_2	Time (min)	6,86	13	22	31	37,13
X_3	Dint (cm)	1.12	1.4	1.8	2.2	2.47

Responses Studied. For this plan, the responses studied are: The dye removal efficiency (Y_1), and the energy consumption (Y_2).

Experiments Matrix. The experimental configuration of three-factor CCD plan with six center repeats is reported in Table 6.

Mathematical Model. The mathematical relationship of the Y response can be approximated by quadratic/(second degree) polynomial equation as shown below:

$$Y = b_0 + b_1x_1 + b_2x_2 + b_3x_3 + b_{12}x_1x_2 + b_{13}x_1x_3 + b_{23}x_2x_3 + b_{11}x_1^2 + b_{22}x_2^2 + b_{33}x_3^2$$
(7)

Where Y is the predicted response, b_0 the constant, b_1, b_2 and b_3 the linear coefficients, b_{12}, b_{13} and b_{23} the coefficients of the cross product, and b_{11}, b_{22} and b_{33} are the quadratic coefficients.

Results of the Experimental Plan
Statistical Validation of Models. Table 7 shows the ANOVA table of the two responses studied.

Table 6. Experimental matrix of the CCD plan

N° Exp	N° rand	Configuration	X_1	X_2	X_3	Y_1 (%)	Y_2 (Wh/m³)	Y_1 Estimated (%)	Y_2 Estimated (Wh/m³)
1	20	---	1000	13	1.4	44.35	1182.75	47.933	1492.82
2	14	--+	1000	13	2.2	13.79	477.99	23.304	640.98
3	16	-+-	1000	31	1.4	57.28	2822.85	67.151	3448.41
4	18	-++	1000	31	2.2	37.31	1226.15	42.522	1520.30
5	17	+--	2000	13	1.4	40.93	2062.08	40.608	2339.17
6	15	+-+	2000	13	2.2	33.75	1063.06	30.629	1487.33
7	7	++-	2000	31	1.4	59.27	4899.97	59.826	5308.21
8	8	+++	2000	31	2.2	45.22	2640.34	49.847	3380.10
9	13	a00	660	22	1.8	43.14	715.71	42.169	1090.41
10	2	A00	2340	22	1.8	47.54	3262.78	42.169	3366
11	5	0a0	1500	6,86	1.8	26.95	397.69	26.934	610.17
12	6	0A0	1500	37,13	1.8	65.58	4164.9	59.255	3846.24
13	3	00a	1500	22	1.12	64.22	4270.53	66.045	4030.49
14	4	00A	1500	22	2.47	36.63	1933	36.943	1692.84
15	19	0	1500	22	1.8	53.33	2452.66	51.494	2228.20
16	11	0	1500	22	1.8	49.39	2281.54	51.494	2228.20
17	9	0	1500	22	1.8	53.85	2183.12	51.494	2228.20
18	10	0	1500	22	1.8	48.23	2035.79	51.494	2228.20
19	12	0	1500	22	1.8	52.1	2033.76	51.494	2228.20
20	1	0	1500	22	1.8	50.98	2300	51.494	2228.20

The ANOVA table reports *p-value* values below 0.05 indicating that the final models are significant at a 95% confidence level.

Table 7. ANOVA table of the two responses studied by the CCD plan

Source of variation	Sum of squares	Ddl	Middle square	F report	*p-Value*
Y_1					
Regression	2752,7318	9	305,859	14,4013	**<0,0001***
Residues	212,3829	10	21,238		
Total	2965,1147	19			
R^2	93%				
Y_2					
Regression	28026651	9	3114072	30,1628	**<0,0001***
Residues	1032420	10	103242		
Total	29059071	19			
R^2	96%				

* Statistically significant at a level of confidence greater than 95% ($p < 0.05$).

The coefficients of determination R^2 are respectively 93 and 96% for the responses Y_1 and Y_2. This shows that only 7 and 4% of the total variations are not described by the models [1–14].

We can conclude that the plan perfectly describes the variations of the two responses. Therefore, it can be used to predict the value of each response at any point in the experimental domain.

Mathematical Models of the Two Responses. Estimation of the coefficients of the models describing the responses studied are reported in Table 8.

Table 8. Estimation of model coefficient for plan

Term	Y_1			Y_2		
	Estimation	Rapport F	Prob. > F	Estimation	Rapport F	Prob. > F
b_0	51.494635		**<0.0001***	2228.2098		**<0.0001***
b_1	2.4778685	3,9481	0.0750	676.53611	60,5447	**<0.0001***
b_2	9.6089227	59,3718	**<0.0001***	962.08801	122,4402	**<0.0001***
b_3	−8.652114	48,1366	**<0.0001***	−694.9879	63,8923	**<0.0001***
b_{12}	−0.83	0,2595	0.6215	253.36188	4,9741	**0.0498***
b_{13}	3.6625	5,0527	**0.0484***	−119.6501	1,1093	0.3170
b_{23}	0.465	0,0814	0.7812	−269.0674	5,6099	**0.0394***
b_{11}	−3.296795	7,3751	**0.0217***	−169.3747	4,0045	0.0733
b_{22}	−2.969758	5,9845	**0.0345***	−66.1194	0,6102	0.4528
b_{33}	−1.498976	1,5247	0.2451	223.96055	7,0015	**0.0245***

* Statistically significant at a level of confidence greater than 95% ($p < 0.05$).

From Table 8, the factors with the most influence on the responses are:

For the Y_1 (dye removal efficiency) response was significantly affected by the synergistic effect of the linear term of time (X_2), with a p-value < 0.0001* and the antagonistic effect of the linear term of the inter-electrode distance (X_3) with a p-value < 0.0001*. While the effect of the conductivity (X_1) is less pronounced (P = 0.075), but its synergistic effect combined with the inter-electrode distance is favourable (X_1X_3) (p-value of 0.0484*) whereas its effect favourable quadratic (X_1^2) (p-value of 0.0217*) but antagonist indicates an improvement of the removal efficiency at very low value.

Although the first order effect shows that the increase in time (X_2) favours the elimination of staining (%), an antagonistic effect of the quadratic term of time (X_2^2) was also observed (p-value of 0.0345*).

Significant factors for the Y_2 (energy consumption) responses were the three linear terms of a synergistic effect for conductivity (X_1) and time (X_2) and antagonistic effect for inter-electrode distance (X_3), with values $p < 0.0001*$. The interaction effect of the variables is considered very significant, except for the interaction between the conductivity and the inter-electrode distance (X_1X_3) (p-value of 0.3170).

Moreover, only the quadratic term of the inter-electrode distance (X_3) which has a significant effect (p-value of 0.0245*) and its effect becomes synergistic.

By taking the statistically significant coefficients from the table, we obtain the following mathematical expressions describing the two responses:

$$Y_1 = 51.494 + 2.477X_1 + 9.608X_2 - 8.652X_3 + 3.662\ X_1X_3 - 3.296\ X_1^2 \\ - 2.969\ X_2^2 \tag{8}$$

$$Y_2 = 2228.209 + 676.536X_1 + 962.088X_2 - 694.987X_3 + 253.361X_1X_2 - 269.067X_2X_3 \\ + 223.96X_4^2 \tag{9}$$

Response Surface Methodology (RSM). Figure 5 illustrates the three-dimensional representations of the two responses. All the surfaces reveal the existence of a peak, which shows that the optimal conditions have been reached, and that we have managed to optimize the responses within the experimental domain.

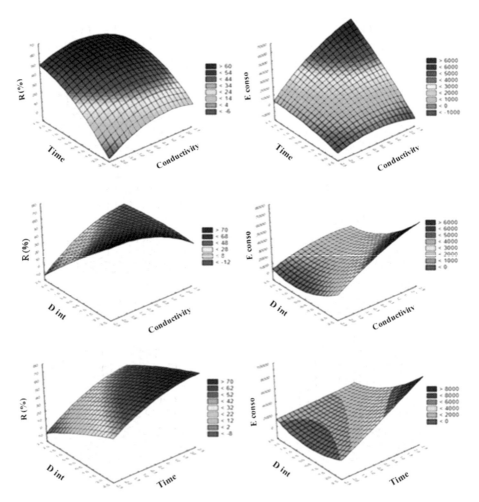

Fig. 5. The 3D response surface contour plot representation of the regression equation for the CCD plan.

Optimization of Parameters. The purpose of this part is to determine the optimal conditions, leading to a maximum elimination of coloring, with lower energy consumption.

The optimization was done through the desirability function stated by Derringer et al. [15].

Indeed, it is more interesting to determine the simultaneous optimal conditions of the two responses studied, through the global desirability function [1–14].

Figure 6 illustrates the overall desirability of the Y_1 and Y_2 responses.

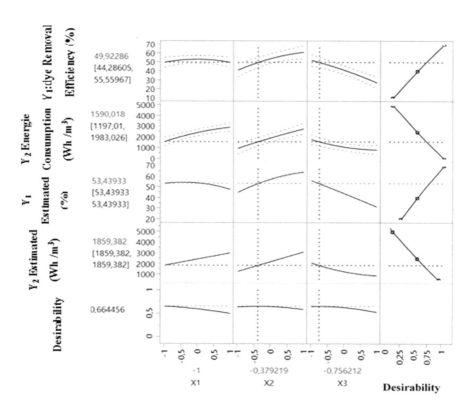

Fig. 6. Graphical representation of the desirability functions associated with the responses: Y_1 and Y_2 for the CCD plan.

Figure 7 illustrates the three-dimensional representations of the global desirability function.

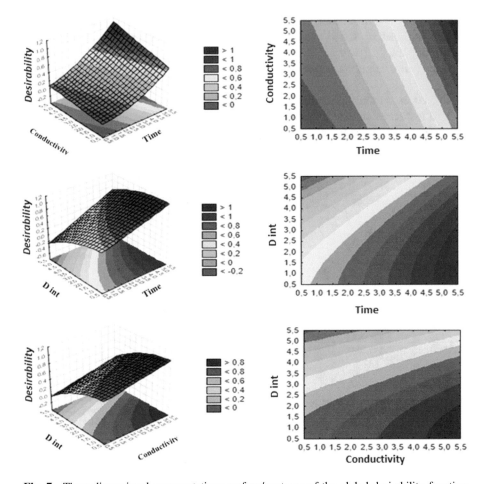

Fig. 7. Three-dimensional representations surface/contours of the global desirability function

Hence, the optimal condition for the treatment of AR35-dyed rejects is given in Table 9.

Under these conditions, the estimated values of dye removal efficiency, and energy consumption for electrocoagulation treatment of AR35 dye are respectively, 53.43% and 1859.38 W h/m^3 and this optimal condition is confirmed experimentally for the two responses (Table 9).

The coded factors were calculated according to the following equation:

$$(X_i - X_{i0})/\Delta X_i \tag{10}$$

Where X_i is the coded value of the independent variable i, X_{i0} is the natural value of the independent variable at the center point, and ΔX_i is the step change value.

Table 9. The conditions of optimization by CCD plan

Variables	Factor	Experimental domain		Optimization results	
		Coded value	Real value	R (%)	Econs
X_1	Electrical conductivity (μS/cm)	-1	1000		
X_2	Time (min)	$-0{,}38$	20,7	52,74	2086,75
X_3	Dint (cm)	$-0{,}75$	1,575		

5 Conclusion

This study clearly demonstrated the applicability of the EC process using the Aluminium Anode for the removal of the AR35 dye. As well as, Response Surface Methodology (RSM), "desirability function" [15] is one of the appropriate methods to optimize the best operating conditions. Satisfactory empirical model equations are developed for two responses, dye removal efficiency, and energy consumption using CCD. The graphical response surface and contour plot are used to locate the optimal point.

Recall that the results obtained have qualitatively very strong similarities to the literature [1–6, 13, 14], as to the role of the main operating parameters (time, distance between electrodes and conductivity).

References

1. Hakizimana, J.N.: Electrocoagulation process in water treatment: a review of electrocoagulation modeling approaches. Desalination **404**, 1–21 (2017)
2. Welham, A.: The theory of dyeing (and the secret of life). J. Soc. Dyers Colour. **116**, 140–143 (2000)
3. Zollinger, H.: Syntheses, Properties and Applications of Organic Dyes and Pigments, Color Chemistry, 2nd edn. V.C.H. Publishers, New York (1991)
4. Peters, R.W.: Wastewater treatment-physical and chemical methods. J. Water Pollut. Control Fed. **57**, 503–517 (1985)
5. Greaves, A.J.: Correlation between the bioelimination of anionic dyes by an activated sewage sludge with molecular structure. Part 1: literature review. JSDC **115**, 363–365 (1999)
6. Daneshvar, N.: Decolorization of basic dye solutions by electrocoagulation: an investigation of the effect of operational parameters. J. Hazard. Mater. **129**, 116–122 (2006)
7. Kobya, M.: Treatment of textile wastewaters by electrocoagulation using iron and aluminum electrodes. J. Hazard. Mater. **B100**, 163–178 (2003)
8. Jiang, J.: Laboratory study of electrocoagulation–flotation for water treatment. Water Res. **36**, 4064–4078 (2002)
9. Bayramoglu, M.: Operating cost analysis of electrocoagulation of textile dye wastewater. Sep. Purif. Technol. **37**, 117–125 (2004)
10. Kobya, M.: Treatment of levafix orange textile dye solution by electrocoagulation. J. Hazard. Mater. **B132**, 183–188 (2006)

11. Kobya, M.: Treatment of potable water containing low concentration of arsenic with electrocoagulation: different connection modes and Fe–Al electrodes. Sep. Purif. Technol. **77**, 283–293 (2011)
12. Saravanan, P.: Biodegradation of phenol and m-cresol in a batch and fed batch operated internal loop airlift bioreactor by indigenous mixed microbial culture predominantly Pseudomonas sp. Bioresour. Technol. **99**, 8553–8558 (2008)
13. Secula, M.S.: An experimental study of Indigo carmine removal from aqueous solution by electrocoagulation. Desalination **277**, 227–235 (2010)
14. Zaroual, Z.: Optimizing the removal of trivalent chromium by electrocoagulation using experimental design. Chem. Eng. J. **148**, 488–495 (2009)
15. Derringer, G.: Simultaneous optimization of several response variables. J. Qual. Technol. **12** (4), 214–219 (1980)

Water Quality of El Hachef River (Region of Tangier-Tetouan-AL Houceima, North West Morocco)

Mouna Elafia[1($\boxtimes$)], Brahim Damnati[1], Hamid Bounouira[2],
Khalid Embarch[2], Hamid Amsil[2], Moussa Bounakhla[2],
Mounia Tahri[2], and Ilyas Aarab[2]

[1] Laboratory of Marine Environment and Natural Hazard,
Faculty of Sciences and Techniques, University Abdelamlek Essaâdi,
BP 416, Tangier, Morocco
mounaelafia@gmail.com
[2] Geochemistry and Chemical Pollution Unit (UGPC),
National Center for Energy, Sciences and Nuclear Techniques (CNESTEN),
BP 1382, El Maâmoura, Morocco

Abstract. This work is part of a chemical and geochemical study of the surface waters and actual sediments of EL Hachef river in the region of Tangier-Tetouan-Al Houceima (NW Morocco). The objectives of this work are determine the physico-chemical parameters, major and trace elements concentrations of waters in order to determine the degree of pollution, to study the geochemical quality and to determine the geochemical inheritance in EL Hachef river area.

The water samples collected were analyzed by the neutron activation analysis technique. Results obtained show that the values of the major and trace elements both in water are representative of those of the upper continental crust as a whole. This confirms the good quality of water and the importance of the geochemical inheritance of EL Hachef river basin.

Keywords: EL Hachef river · Water · Major elements · Trace elements · Pollution · Quality · Morocco

1 Introduction

Hundreds of pollutants are dumped every day into the environment among them, major and traces elements that are considered serious pollutants of the aquatic environment [1]. Several major and trace elements are found in the aquatic environment by human action, by atmospheric transport and as a result of erosion [2]. If organic pollutants may degrade, metal pollutants persist in the aquatic environment. Therefore, the study of their mobility and transfer is of prime importance. Generally, contamination of aquatic ecosystems by major and trace elements can be confirmed in water, sediments and organisms [3]. Major and trace elements could be dissolved in water and in the interstitial water of the sedimentary column or fixed on the sediment particles or suspended in water.

© Springer Nature Switzerland AG 2019
M. Ezziyani (Ed.): AI2SD 2018, AISC 913, pp. 61–72, 2019.
https://doi.org/10.1007/978-3-030-11881-5_6

The distribution of major and trace elements in sediment samples determines their behaviors in the environment. The major and trace elements resulting from the anthropic inputs behave differently from the elements that naturally exist in the environment. Once introduced into the sediments, they will associate with different constituents of the latter in various chemical forms. This distribution depends on the intrinsic characteristics of the element (initial chemical form, reactivity, etc.) and the characteristics of the area (pH, redox potential, temperature, etc.).

In Morocco, the assessment of water and sediment quality is carried out mainly for the main hydrographic networks such as those of Sebou [4], Bouregreg [5]. The diagnosis and analysis that have been developed by a large number of researchers make it possible to draw conclusions:

In Sebou river, results of analysis of sediment samples show contamination in some trace elements evaluated, namely Cr, Mn, As, Ti, Sb, Zn, V, Co and Fe on the studied sampling sites. The use of enrichment factor. allowed us to distinguish elements that are enriched such as As and Cr, and the less rich elements, which are Ti, Sb and Zn. These metals could be natural or anthropogenic from discharges sewage in rural areas, including the leaching of garbage or uncontrolled irrigation water rich in fertilizers in the surrounding farmlands.

In Bouregreg river, Results obtained show typical metal enrichments estimated through the enrichment factor relatively to the upper continental crust composition. A general diffuse origin of heavy metals was observed and some local inputs may occur linked urban wastes (garbage dump in the Aguelmous area with Sb and As enrichment).

The Mediterranean and Tangier basins have been little studied. In this context, we would like to contribute with this study whose the main objective is to assess the water and actual sediment quality of EL Hachef river, based on physicochemical approaches and the determination of concentrations of major and trace elements in water and sediments by the use of the neutron activation analysis technique.

2 Materials and Methods

El Hachef river basin belongs to the region of Tangier-Tetouan-Al Houceima, in the North-West of Morocco. It is bordered on the north by the Strait of Gibraltar, on the west by the Atlantic Ocean, on the south by the Gharb-Chrarda-Beni Hssen region and on the east by the Taza-Al Hoceima-Taounate region.

The study area is part of the Rif region, which is a segment of the Alpine chains of the Mediterranean. The western part of the Rif, which is the subject of this study, is subdivided into three major structural and paleogeographic domains [6].

To assess the degree of metallic contamination of EL Hachef river, five samples of water were collected in February 2014. The process of water sampling was also preceded by measurement of physico-chemical parameters, temperature, pH, dissolved oxygen and conductivity. Subsequently, the water samples were taken from ½ liter polyethylene bottles previously washed with dilute nitric acid and then rinsed with distilled water. Each flask is filled individually after rinsing 3 times with the sample to be taken (Fig. 1).

Fig. 1. Location of sampling stations in the EL Hachef river drainage basin.

3 Results and Discussion

3.1 Results of Physicochemical Parameter

Temperature. The temperature of water is a parameter of major importance in the life of aquatic ecosystems. It has an influence on several physical, chemical and biological processes. In general, the water temperature is a function of the ambient temperature. Measured values of temperature at EL Hachef river range between 13.2 °C, 16 °C and 14.7 °C upstream, intermediate and downstream, respectively (Fig. 2), indicating excellent water quality according to the surface water quality grid (source: National Council of the Environment).

For spatial variations, no significant differences were noted. The discrepancies recorded are in fact only due to the daily time difference between the different sampling points. This is the reason why the spatial profile does not show a significant difference between the different stations. The average of 14.36 °C depends on local conditions: climate, duration of sunshine and depth.

Hydrogen Potential (pH). Measurements of pH reveal that the water of EL Hachef river is slightly alkaline. Overall, pH values decrease from upstream to downstream. This is probably due to the nature of the lands crossed (Region rich in limestone's) and to the different domestic discharges (presence of rural communes near the river as Dar chaoui, El Hajra, El Khoucha, Ain ben amar). The observed values are between 7.37

and 8.15 (Fig. 3). This translates to excellent water quality since the pH is still between 6.5 and 8.5 according to the National Environment Council quality grid.

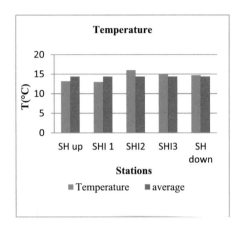

Fig. 2. Water temperature variation at EL Hachef river.

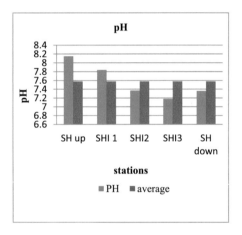

Fig. 3. Water pH variation measured in situ at the EL Hachef river.

Dissolved Oxygen. Oxygen is one of the most useful parameters for water and is an excellent indicator of quality. Its presence in surface waters plays a preponderant role in the self-purification and maintenance of aquatic life.

Generally, waters of EL Hachef river are less oxygenated. They demonstrate relatively higher contents upstream than 7.2 mg/l compared to the ones measured downstream 5.1 mg/l. This spatial evolution of the dissolved oxygen content at the level of EL Hachef river shows a significant decrease in the oxygenation of the environment and especially in the three stations SHI 2, SHI 3 and SHdown (Fig. 4).

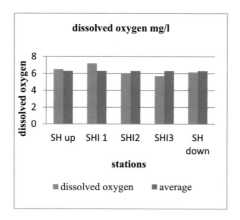

Fig. 4. Variation of dissolved oxygen measured at EL Hachef river.

This data implies that water quality is good since this data is always reliable to the standard [5–7 mg/l] according to the surface water quality grid (source: National Council of the Environment).

Electrical Conductivity. Water conductivity is an indicator of changes in the composition of materials and their overall concentration. It provides information about the degree of overall mineralization of surface waters.

Results presented in our study show that the recorded values remain generally varied along EL Hachef river with values that vary between 420 μs/cm and 490 μs/cm except for the intermediate station SHI 2, which reaches 680 μs/cm (Fig. 5). These values can be explained by the natural mineralization linked to the nature of the limestone soils of the watershed. Overall, these waters remain classified as excellent quality watercourses since conductivity is always below the 750 μs/cm standard according to the surface water quality grid (source: National Environmental Council).

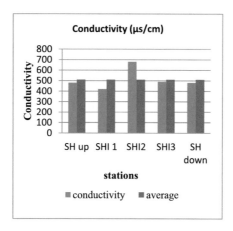

Fig. 5. Water conductivity variation measured in situ at EL Hachef river.

3.2   Concentrations of Major and Trace Elements in EL Hachef River

Major Elements. The upstream-downstream evolution of the content of major elements Na, Mg, Al, K, Ca, Ti, Mn and Fe in the water of EL Hachef river shows a very clear homogeneity in all the stations.

The evolution of the concentrations of Mg, Al, Ti, and Mn is almost similar. These elements have low concentrations in most study stations. They do not exceed 15.35 mg/kg, 25.46 mg/kg, 5.44 mg/kg and 0.2 mg/kg, respectively. Concentrations of the other major elements, Na, K, Ca and Fe are more or less higher. They do not exceed 135.5 mg/kg, 201.5 mg/kg, 116.3 mg/kg and 238.1 mg/kg respectively.

The importance of concentration of these elements throughout the basin implies a multiple origin (agricultural or natural zone) as well as the leaching of waste from the villages near the river contributes also to the enrichment of these elements. On the other hand, the erosion of the soil would be responsible for this enrichment in these major elements since the study area is considered as a zone of important agricultural activity.

Trace Element. Concentrations of the trace elements Cl, Sc, V, Cr, Co, Zn, As, Se, Rb, Sr, Zr, In, Sb, I, Cs, Ba, and Hf, found in the water samples of EL Hachef river, have an almost similar distribution along the river. However, Cl, and Zr abound in all stations at high levels of 164.4 mg/kg and 162.4 mg/kg, respectively. Levels of Cl, Zn and Zr remain higher compared to other elements not exceeding 4.60 mg/kg. These levels can be explained by the discharges of waste water from rural areas.

Rare Earth Elements (REE). The spatial evolution of the rare earth elements presents small variations. A slight increase in the contents from upstream to downstream, where the minimum concentrations recorded at the SH_{up} station and the maximum ones at the SH_{down} station. It should also be noted that the concentrations determined in water samples of the river generally remain low in rare earths La, Ce, Nd, Sm, Eu, Tb, Dy, Tm and Yb regarding to major and trace elements. This decrease can be due to the dilution effect generated by the tributaries, underlining the absence of industrial activities in the study area.

Results of major and trace elements in water samples of EL Hachef river are shown in Table 1.

Table 1. Concentrations of major and trace elements in water samples of El Hachef river/ (mg/kg).

	SH_{up}	SHI 1	SHI2	SHI3	SH_{down}
Na	110	119	135	111	114
Mg	9.941	13.56	15.35	15.08	13.67
Al	8.9	6.1	9.9	6.3	25.4
Cl	123	136	164	128	130
K	171.2	192	191.3	183	201.5
Ca	71.7	116.3	106.3	98	80.29
Sc	0.008	0.02	0.01	0.014	0.02

(continued)

Table 1. (*continued*)

	SH$_{up}$	SHI 1	SHI2	SHI3	SH$_{down}$
Ti	2.816	3.868	5.448	5.302	4.196
V	0.034	0.025	0.044	0.04	0.04
Cr	1.073	1.44	1.58	1.87	2.7
Mn	0.067	0.025	0.15	0.05	0.2
Fe	99.27	137.6	133	172.7	238.1
Co	0.18	0.24	0.21	0.3	0.43
Zn	85	24.6	1085	42	53.2
As	0.42	0.6	0.61	0.89	0.72
Se	0.56	0.78	0.8	0.9	1.29
Br	0.15	0.56	0.51	0.32	0.66
Rb	5.35	7.47	7.12	9.51	13.2
Sr	51.42	73.26	72.02	87.25	125.7
Zr	65.73	89.15	99.18	111.5	162.4
In	0.003	0.0036	0.003	0.003	0.014
Sb	0.13	0.12	0.12	0.16	0.13
I	0.113	0.139	0.148	0.143	0.142
Cs	0.61	0.85	0.84	1.06	0.94
Ba	3.817	4.592	4.609	4.379	4.556
La	3.31	3.82	3.75	3.64	3.9
Ce	2.07	2.88	2.67	3.49	3.07
Nd	2.45	3.31	3.48	4.15	5.63
Sm	0.014	0.019	0.018	0.02	0.02
Eu	0.017	0.02	0.02	0.018	0.019
Tb	0.04	0.05	0.06	0.07	0.1
Dy	0.008	0.021	0.02	0.02	0.02
Tm	0.19	0.2	0.26	0.31	0.42
Yb	0.39	0.53	0.37	0.48	0.56
Hf	0.08	0.11	0.117	0.13	0.2
Ta	0.07	0.078	0.08	0.07	0.16
Th	0.08	0.11	0.12	0.14	0.21
U	0.21	0.2	0.26	0.15	0.23

3.3   Spectra of Elements Normalized to the Continental Crust

In order to check the typology of the El Hachef river basin regarding to the upper continental crust, the contents of major and trace elements analyzed have been normalized to the upper continental crust content (UCC) given by [7].

Diagrams of mean concentrations normalized to the upper continental crust are superimposable. Spectra of normalized concentrations in water samples (Fig. 6), have the same shape regardless of the station, with the exception of Zn at station SHI 2.

Normalization to the upper continental crust shows negative anomalies in Na, Mg, Mn, Fe, Co, Rb, Sr, In, Cs, Ba, La, Ce, Nd, Sm, Eu, Tb, Dy, Yb, Hf, Ta, Th and U for all sampling sites. Elements showing negative anomalies are the most easily mobilized during the alteration of the rocks [8]. The other elements Zn, As, Se, Zr, Sb and Tm seem to have the same variations both in the basin and in the upper continental crust. Their reports of concentrations relative to the upper continental crust are almost equal to the unit.

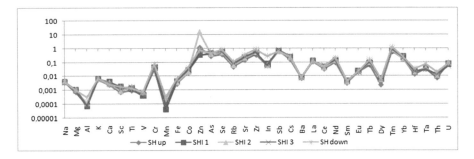

Fig. 6. Normalization diagram of major and trace elements in El Hachef river water. samples in relation to the upper continental crust.

3.4   Enrichment Factor

The concept of enrichment factor is based on establishment a quantitative assessment of the pollution degree for each element in relation to the upper continental crust [7] and to determine the origin (natural and/or anthropogenic) of these elements in aquatic environments [9]. The enrichment factor is calculated as follows:

$$\left(FE(i) = \{[C_i]/[C_{Fe}]\}_{samp} / \{[C_i]/[C_{Fe}]\}_{UCC} \right) \tag{1}$$

With $\{[C_i]/[C_{Fe}]\}_{samp}$: Ratio between concentrations of element i and the reference element (Fe) in the sample.

$\{[C_i]/[C_{Fe}]\}_{UCC}$: Ratio between concentrations of element i, and the reference element (Fe) in the upper continental crust.

Normalization with respect to Fe is chosen because their concentrations show very limited variations across the studied river basin.

The Case of Water Samples from El Hachef River. Enrichment factor (EF) values calculated for major and trace element in water samples of El Hachef river (Table 2) range from 0.02 to 150.84. Limits at 10 and 100 are conventionally used [10]. We can thus separate the depleted elements (FE < 10), moderately enriched (10 < FE < 100) and highly enriched (FE > 100). Three groups of elements were distinguished:

- The depleted elements: Na, Mg, Al, k, Ca, Sc, Ti, V, Mn, Fe, Ba, Ce, Sm, Dy, Hf and Th.
- The moderately enriched elements Cr, Co, As, Se, Sr, Zr, In, Sb, Cs, La, Nd, Tb, Yb and U.
- The other elements Zn and Tm are therefore the highly enriched elements.

Table 2. Enrichment factor in water samples of El Hachef river.

	SH_{up}	SHI 1	SHI 2	SHI 3	SH_{down}
Na	1.35380167	1.05711816	1.23907742	0.78663203	0.58486668
Mg	0.2662841	0.26204413	0.30689508	0.23218929	0.15266599
Al	0.03910636	0.01953684	0.03243181	0.01593699	0.04658935
K	2.14504222	1.73824487	1.78901275	1.31798012	1.05260581
Ca	0.85063418	0.99485722	0.94076489	0.66800164	0.39691868
Sc	0.25714992	0.46379493	0.23991798	0.25867242	0.268031
Ti	0.33189483	0.32889244	0.47926015	0.35919745	0.20618732
V	0.20036265	0.10628634	0.19353383	0.13549508	0.10564889
Cr	10.8397876	10.4950166	11.9136412	10.8589627	11.3721726
Mn	0.03845769	0.01035257	0.06426374	0.01649697	0.04786268
Fe	1	1	1	1	1
Co	6.36446056	6.12209302	5.54210526	6.09727852	6.33893322
Zn	423.351168	88.4182566	4032.9874	121.831067	110.645841
As	99.0027199	102.034884	107.323308	120.59062	70.7601848
Se	132.003626	132.645349	140.75188	121.94557	126.778664
Rb	16.8898135	17.0134058	16.7771214	17.2574758	17.3741525
Sr	51.946121	53.393397	54.3050913	50.665481	52.9437811
Zr	122.320466	119.689604	137.760902	119.271325	126.002785
In	21.2148685	18.3662791	15.8345865	12.194557	41.2767745
Sb	229.827743	153.052326	158.345865	162.594094	95.8210836
Cs	58.2931073	58.6011157	59.9146515	58.2262633	37.4518996
Ba	2.45385313	2.1297463	2.21156391	1.61818077	1.22114925
La	39.011786	32.4811047	32.9887218	24.6601042	19.1642167
Ce	11.4361401	11.4789244	11.0099859	11.083074	7.0714117
Nd	33.318223	32.474564	35.3233083	32.4406485	31.9214616
Sm	1.10003022	1.07703488	1.0556391	0.90330052	0.6551869
Eu	6.8305448	5.79743658	5.99794942	4.15723535	3.18286816
Tb	22.0988214	19.9286882	24.7415414	22.2296613	23.0339143
Dy	0.80818547	1.53052326	1.50805585	1.16138638	0.84238315
Tm	203.577021	154.598309	207.928913	190.924883	187.621702
Yb	62.6802934	61.4528277	44.3848257	44.3438438	37.5243404
Hf	4.87698127	4.83786087	5.32369717	4.55543797	5.08334661
Ta	11.2503091	9.04400106	9.59671907	6.46681055	10.7212401
Th	2.64359733	2.62239187	2.95973579	2.6592492	2.89323185
U	26.5185857	18.220515	24.5059076	10.8879974	12.1092578

3.5   Principal Component Analysis (PCA)

In this work, the PCA was used to study thoroughly correlations between various determined elements and the sampling points in El Hachef river both for water samples.

The statistical treatment of El Hachef river water is carried out on 5 observations and 35 variables (Na, Mg, Al, Cl, K, Ca, Sc, Ti, V, Cr, Mn, Fe, Co, Zn, R, Sb, Zr, In, Sb, I, Cs, Ba, La, Ce, Nd, Sm, Eu, Tb, Dy, Tm, Yb, Hf, Ta, Th and U).

Case of Water Samples of El Hachef River. Major components 1 and 2 account for 77.03% of overall inertia; 56.82% of the information available in the input data table is explained by the first main component 1 and 20.21% is explained by the second main component.

Relationship Between Main Components 1 and 2 and the Variables. In the plane formed by main components 1 and 2 (Fig. 7), the variables most correlated to the first principal component (PC 1) are: Al, K, Sc, Cr, Fe, Co, As, Se, Rb, Sr, Zr. Among these variables, those most contributing to the formation of PC 1 are: Cr, Fe, Fe, Fe, Fe, Fe, Tb, Co, Se, Rb, Sr, Zr, Ce, Nd, Tb, Tm, Hf and Th, which are strongly correlated to each other since they define eigenvectors in the same direction. Correlations between these variables are stronger as they are positioned at the ends of the axis defined by the PC 1.

On the axis defined by the second principal component (PC 2), the variables most correlated to this main component are Na, Mg, Cl, Ca, Ti, Zn and Eu. The other variables Mn, Sb, U and V are close to the center of the factor plane; their correlation is certainly not very strong.

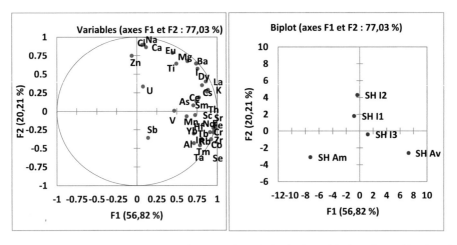

Fig. 7. Distribution of variables (a) and observations (b) in the factorial plane formed by the PC 1 and PC2 in water samples of El Hachef river.

The elements Na, Mg, Cl, K, Ca, Ti, V, As, Cs, Ba, La, Sm, Eu, Dy and U are on the upper right quarter of the factorial plane, with thus "positive" links with PC 1 and

PC 2. On the other hand, because of their positions at the center of the factorial plane, we can deduce that the U and V are not very well explained by the main components 1 and 2. The variables Al, Sc, Ti, Cr, Mn, Fe, Co, Se, Rb, Sr, Zr, In, Sb, I, Ce, Nd, Tb, Tm, Yb, Hf, Ta and Th are on the lower right quarter of the factorial plane. Therefore, these elements have "positive" links with PC 1 and "negative" links with PC 2 and have stronger links with PC 1 than with PC 2. Mn and Sb are not well explained by the main components 1 and 2 since they were at the center of the factorial plane.

Zinc occupies the upper left quarter of the factorial plane; this element consequently has "negative" links with CP1 and "positive" links with PC2.

4   Discussion

Results obtained from the various physicochemical parameters of water samples of El Hachef river allowed us to classify it among rivers of good to excellent quality which meet the values set by the grid of the quality of surface water in Morocco.

The choice of the Neutron Activation Analysis (NAA) method proved to be a good idea. This method allowed us to detect a large number of elements (Mg, Al, Cl, K, Ca, Sc, Ti, V, Cr, Mn, Fe, Co, Zn, Zr, In, Sb, I, Cs, Ba, La, Ce, Nd, Sm, Eu, Tb, Dy, Tm, Yb, Hf, Ta, Th and U) found in water and actual sediment samples of El Hachef river. As a result of standardization results compared to the upper continental crust, the normalized spectra of the major and trace elements in El Hachef river basin are representative of those of the upper continental crust as a whole.

Hence, the normalization of the elements detected by NAA in El Hachef river concludes:

- The importance of the geochemical heritage of the geological environment of El Hachef river basin.
- The geochemical homogeneity of the basin which demonstrate a geochemical inheritance during crustal accretion (magmatism-metamorphism-sedimentation).

5   Conclusion

Results obtained of El Hachef river allowed us to classify it among rivers of good to excellent quality which meets to the values set by the grid of the quality of surface water in Morocco.

Acknowledgements. This work was only possible thanks to the participation of the Laboratory of analysis by neutron activation (CNESTEN-Rabat-Morocco) and to the members of EMRN-FST-Tangier (Abd Elmalek Essaâdi University).

References

1. Harte, J., Holdren, C., Schneider, R., Shirley, C.: Toxics A to Z, A guide to everyday pollution hazards. 0-520-07224-3. 478 (1991)
2. Veena, B., Radhakrishnan, C.K., Chacko, J.: Heavy metal induced biochemical effects in an estuarine teleost. Indian J. Marine Sci. **26**, 74–78 (1997)
3. Förstner, U., Wittmann, G.T.W.: Metal pollution in aquatic environment. 0-540-12856-4.486 (1983)
4. Foudeil, S., Bounouira, H., Embarch, K., Amsil, H., Bounakhla, M., Ait Lyazidi, S., Benyaich, F., Haddad, M.: Evaluation de la pollution en métaux lourds dans l'oued Sebou (Maroc) 130906, 2111-4706 (2013)
5. Bounouira, H.: Etude des qualités chimiques et géochimiques du bassin versant du Bouregreg. Thèse (doctorat) 00726475295 (2007)
6. Durand Delga, M.: Données actuelles sur la structure du rif. Mem. Hrs. Serv. Soc. Geol. Fr. livre. mem. p. Fallot **1**, 399–422 (1961)
7. Taylor, S.R., McLennan, S.M.: The geochemical evolution of the continental crust. Rev. Geophys. **33**, 241–265 (1995)
8. Nesbitt, H.W., Markovics, G., Price, R.C.: Chemical processes affecting alkalis and alkali earths during continental weathering. Geochim. Cosmochim. Acta **44**, 1659–1666 (1980)
9. Tam, N., Yao, M.: Normalization and heavy metal contamination in mangrove sediments. Sci. Total Environ. **216**, 33–39 (1998)
10. Roy, S., Négrel, P.: A Pb and trace element study of rainwater from the Massif Central (France). Sci. Total Environ. **277**, 225–239 (2001)

Optimization of Solid Waste Composting: A Literature Review and Perspective for Fast Composting

Manale Zouitina[(⊠)], Khadija Echarrafi, Ibtisam El Hassani, and Mounia El Haji

ENSEM, Hassan II University, Casablanca, Morocco
m.zouitina@ensem.ac.ma

Abstract. Composting may be considered as a simple process. However, it requires specific aerobic conditions and control of its important parameters (oxygen, temperature, moisture content). Composting technology is an effective means for the recovery of organic waste and the amendment of soils. This aerobic process involves various microorganisms for the decomposition of organic matter and its transformation into a product to sustainably improve soil fertility.

The trend towards more efficient and fast methods of compost production requires a thorough and understanding of the process as well as its main parameters in order to develop a product which is not only beneficial in terms of quality and maturity but also can be obtained in a short time to satisfy agriculture's needs.

This literature review aims to present the composting process as addressed in different studies and the optimization methods that have provided promising results in reducing time composting, it also gives a perspective for fast and smart composting strategy.

Keywords: Composting · Organic matter · Green waste · Optimization · Maturation

1 Introduction

Waste treatment is actually one of the most important problems. The constant increase in the volume of waste produced requires fast and effective processing means generating little or no nuisance.

Recently, Morocco has begun to invest more and more in sustainable development, which justifies its strategic orientation towards green projects. Composting is considered as an important component of waste valorization in terms of reusing the organic waste for agricultural purposes. It consists on degrading the biomass and its transformation to a bio fertilizer useful for soil amendment.

This review study represents composting process, major factors that affect compost maturity and explores the various techniques used in the optimization of the composting process in order to open research perspectives in terms of intelligent composting able to self-adjust, as a promising strategy of improvement to both efficiency and quality of fast composting.

© Springer Nature Switzerland AG 2019
M. Ezziyyani (Ed.): AI2SD 2018, AISC 913, pp. 73–83, 2019.
https://doi.org/10.1007/978-3-030-11881-5_7

2 Composting Process

Composting is the biological decomposition and stabilization of organic substrates under conditions that allow development of thermophilic temperatures as a result of biologically produced heat, to produce a final product that is stable, free of pathogens and plants seeds, and can be beneficially applied to land [1].

This process can be also defined as a spontaneous degradation of the organic matter by mixed microbial population in a warm, moist and aerobic environment given the right conditions. Compost quality is related to an important parameter which is compost maturity [2] (Fig. 1).

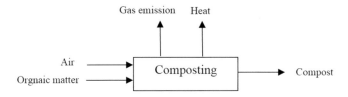

Fig. 1. Conceptual diagram of composting process

It's a solid-waste process that requires aerobic conditions. The organic matter is degraded by different micro-organisms [3]. The basic process of composting is exothermic oxidation of the bio waste organic matter, transformed by chemo-heterotrophic microorganisms through the three main phases of composting (Fig. 2).

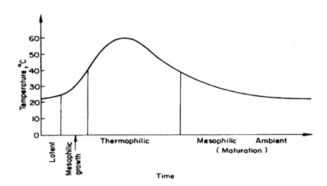

Fig. 2. Evolution of temperature profile during the composting process time

The mesophilic phase, the thermophilic phase and the maturation or curing phase [4, 5]. In the mesophilic phase, the aerobic degradation of organic matter, raises the temperature to thermophilic values which characterizes the thermophilic step then the temperature decreases in the cooling or maturation phase [5–7].

The curve bellow shows the evolution of temperature profile during the composting process time [8].

Biodegradable organic wastes that can be managed by composting are generated in a diverse range of production sectors, including primary production systems, industries, and services such as: Municipal organic waste, commercial food waste, forestry, agriculture, sewage sludge, food and beverage industry.

Sources of organic waste of composting [9] are listed in the Table 1 below:

Table 1. Compost inputs.

Organic waste	Examples
Commercial food waste	Waste from markets, catering…
Food and beverage industry	Waste from breweries and malt houses, fruit and vegetable production, meat production…
Municipal organic waste	Biodegradable fraction of municipal organic waste
Sewage sludge	Sludge from biological treatment of municipal Wastewater
Agriculture	Solid and liquid slurry, straw residues, sugar beet and potato haulm, vegetables…
Forestry	Wood residues, leaves, bark

3 Composting Systems

Controlling the composting process requires the interdependent factors of water disposal, heat production, temperature and ventilation. These links were taken into account different composting systems, using equipment to provide oxygen to the composting material by aeration. There are various types of composting based on reactors, aeration rates, and temperature changes. These include windrow composting, aerated static pile (uncovered and covered), in-vessel and vermi-composting.

Composting systems can be classified into [9]:

3.1 Open Composting Systems

Static Pile: natural ventilation (passive aeration with chimney effect); forced aeration with suction or blowing in windrows or in silos;

Turned Pile: natural ventilation and mechanical turning, in corridor silos, in windrows, or in piles.

3.2 Closed Composting Systems Functioning Continuously or Periodically with Forced Aeration (In-vessels or Reactor Composting)

Vertical flow reactors: moving agitated bed reactors; and moving packed bed systems;

Horizontal flow reactors: rotary drums; agitated bed bins and agitated solids bed (Table 2).

Table 2. Comparison between different composting systems

Composting method	Advantages	Limitations
Windrow system	Time: 12–20 weeks Cost: Medium	Slow process Low nuisance control Require large area
Static pile	Time: 10–13 weeks Cost: Low	Slow process Low nuisance control Require large area
In-vessel	Rapid composting process Better nuisance control Multiple turnovers of site footprint	High capital and operating cost
Channels(enclosed)	Rapid composting process Better nuisance control Multiple turnovers of site footprint	High cost

4   Composting Parameters

The whole process of composting is influenced by different factors such as pH, Temperature, Aeration, C/N ration, Electrical conductivity, moisture content, particle size and the frequency of turning.

4.1   pH Level

The pH is an important factor in composting [10] because it affects microbial activities during the composting process. [11] noted that high pH level provoked can increase the NH3/NH4 ratio. In [12] it was delineated that aerobic composting presented higher pH than anaerobic process probably caused by the higher potassium release.

4.2   Turning Frequency

Turning of the compost material is the most common procedure to aerate it because it facilitates the degradation of the organic matter and prepares the conditions for the microbial activity and thus, results in gas emission [13].

Improved aeration at the start of the composting process can accelerate the time of compost maturation but the excess of turning the waste can affect the vital components of composting [14]. Consequently the turning regime must be pursued to keep the relevant nutrients or to obtain other specific objectives such as pathogen reduction [15]. It was reported in [16] that there is a strong link between turning frequency and other physico-chemical parameters which could reveal the compost quality [17].

4.3   Temperature

One of the most important factors of composting process is temperature [10, 18, 19]. It was observed in [20] that composting is an exothermic process that depends on the initial temperature and biodegradability of the substrates. Temperature is also considered as a function of the process as stated by [21]. [19] averred that the accelerated biodegradation organic matter by microorganisms caused an increase temperature during composting.

4.4   C/N Ratio

During composting, microorganisms decompose organic compounds to obtain energy for metabolism and acquire nutrients (such as P, N, K) to sustain their populations [22]. C, P, N and K are the major nutrients needed by the microorganisms associated with composting [23]. However, C and N are the most important: C is used as energy source while N is used for building cell structure [22, 24, 25] reported that when the amount of N is limiting, microbial growth decreases and it results slow decomposition of the compound C.

4.5   Electrical Conductivity

Electrical conductivity was defined in [26] as a numerical expression of conduction of electrical current by an aqueous solution. Electrical conductivity indicates the salinity of an organic amendment [27]. Within the composting process concentration of salts raises due to the decomposition of complex organic matter [28].

4.6   Particle Size

The particle-size distribution of the final compost is essential because it identifies gas and water exchange especially water-holding capacity [29].

4.7   Aeration

Aeration is a fundamental factor that affects composting process [10]. Basically, composting is an aerobic process, in which the oxygen is consumed and gaseous H_2O and CO_2 are emitted [14]. Efficiency of the composting process is highly dependent on the O_2 level, because the composting is directly related to microbial population dynamics [30].

4.8   Moisture Content

Moisture content is a central factor in the composting process. It influences the oxygen consumption, microbial activity, free air space, and temperature of the process [31]. For [2], the optimal for effective composting is related to the waste type or form. They stated that the feedstock moisture content should be at 50%–60%.

The main characteristics of compost quality and some patterns of composting at the initial stage are listed and tabulated below (Table 3).

Table 3. Optimum compost characteristics

Major parameters of composted waste	Responsible range	Optimum characteristics for fast composting
C/N ratio	[20, 40]	[25, 30]
pH	[5.5, 9]	[6.5, 8]
Oxygen concentration %	>5%	>>5%
Particle size (diameter, mm)	[5, 15]	Varies
Moisture content %	[40%, 65%]	[50%, 60%]
Electrical conductivity (mS/cm)	[3, 10]	
Temperature °C	[40, 65]	[55, 60]

5   Optimization Composting Methods

5.1   Thermophiles Inoculation

The composting process could be accelerated by adding thermophilic bacterial species isolated from a composting aquaculture waste mixture. The composting period was shortened by about a third when the compost material was inoculated with these bacteria compared with inoculated compost [32].

The inoculation of thermophilic microbes had been suggested as a potential approach for accelerating the composting process [33]. Some other studies have demonstrated the efficacy of using thermophiles inoculation to accelerate composting of organic waste [34–37].

5.2   Temperature Control

Temperature control strategy to enhance the activity of yeast inoculated into compost raw material for accelerated composting was proved by [38] due to the microorganisms contained in the yeast which are able to degrade organic acids of the organic matter. Thus, it was showed that temperature control enhances the effects of microbial inoculation into composts.

The study made in [39] suggested Continuous Thermophilic Compost as an effective method for rapid degradation and maturation of organic municipal solid wastes [40] in which the compost materials was incubated in high temperature during the whole composting process.

The authors found that this method could evidently reduce the composting cycle to 14 days without a compost product quality degradation and suggested a method for rapid degradation and maturation of municipal solid waste as follows: heat rapidly the composting materials to 50 C and keep the surrounding at that temperature point during the whole composting process, provide suitable amount of air and water and turn over the pile daily.

5.3 pH Control

The composting process could be restricted by a combination of low pH and high temperature in the mesophilic composting phase [41]. The degradation rate of organic matter in the pH controlled experiment was faster than that without, moreover analogous results can be obtained by liming [42] and or by the use of special microbial inoculants [34] There may also be other methods to increase the pH value in the material, for example by reuse of composted material at high pH together with the other fresh raw ingredients of the compost material.

5.4 Using Additives

Biochar is suggested as a beneficial additive and bulking agent in composting [43, 44]. The use of biochar produced from lawn waste had significant results on mineralization rates and the degradation of food waste compost and ameliorates physiochemical properties of the final compost. Tests showed that biochar amended compost mixtures rapidly achieved the thermophilic temperature, increased the organic matter degradation concentration of NH4 and NO3 [45].

Another study [46] investigated the changes in physical and chemical characteristics during the two-stage composting [29] of green waste [47] with or without adding seaweed and bentonite to the compost material. Addition of 35% seaweed and 4.5% bentonite produced the highest quality and the most mature compost in only 21 days.

The combined addition of bean dregs and crab shell powder as additives during the two-stage composting of green waste leaded to a highest quality compost product in only 22 days [48].

The Table 4 below summarizes some strategies of fast composting as represented in the literature:

Table 4. Fast composting strategies

Methods	Material used	Composting duration (days)	References
Using additives	Elemental sulphur	120	Bustamante et al. [49]
	Green waste and bio solids	60	Belyaeva et al. [50]
	Pic manure	48	Wang et al. [51]
	Fish pound sediment and rock Phosphate	22	Zhang and Sun et al. [52]
	Biochar	22	Zhang and Sun [29, 45], 45 [53] Godlewska et al. [54] Waqas et al. [45] Xiao et al. [55]

(*continued*)

Table 4. (*continued*)

Methods	Material used	Composting duration (days)	References
	Jaggery	21	Gabhane et al. [56]
	Seaweeds and bentonite	21	Clarke et al. (2018) Zhang et al. [46]
Thermophilic bacteria inoculation	Pleurotus sajrorcaju, Trichoderma harzianum, Aspergillus niger Azotobacter Pichiakudriavzevii RB1	14	Chen et al. [32], hu-HsienTsai et al. (2017), Sundberg et al. (2004) [35], Nakasaki et al. [57]

6 Conclusion and Perspectives

Different methods of optimizing the final compost quality in terms of shorting the maturation time and improving its quality by modifying controlling and monitoring some principal parameters of the process. However, there is no researches that treated the fast compost with an aspect related to smart composting equipment which aims to accelerate the total composting duration in terms of hours instead of days without affecting its final quality. This could be an avenue for our further researches.

References

1. Haug, R.T.: The Practical Handbook of Compost Engineering. CRC Press, Boca Raton (1993)
2. Bernal, M.P., Alburquerque, J.A., Moral, R.: Composting of animal manures and chemical criteria for compost maturity assessment. A review. Bioresour. Technol. **100**(22), 5444–5453 (2009)
3. Ryckeboer, J., et al.: A survey of bacteria and fungi occurring during composting and self-heating processes. Ann. Microbiol. **53**(4), 349–410 (2003)
4. Compost: production, use and impact on carbon and nitrogen cycles/ by Maria Pilar Bernal. Book online read or download. http://libisbn.ru/Compost–production-use-and-impact-on-carbon-and-nitrogen-cycles–or–cby-Maria-Pilar-Bernal/1/bgahfag. Accessed 03 Apr 2018
5. Keener, H.M., Elwell, D.L., Monnin, M.J.: Procedures and equations for sizing of structures and windrows for composting animal mortalities. ResearchGate (2000). https://www.researchgate.net/publication/289133805_Procedures_and_equations_for_sizing_of_structures_and_windrows_for_composting_animal_mortalities. Accessed 03 Apr 2018
6. Bernal, M.P.: Compost: production, use and impact on carbon and nitrogen cycles (2008)
7. Oudart, D., Paul, E., Robin, P., Paillat, J.M.: Modeling organic matter stabilization during windrow composting of livestock effluents. Environ. Technol. **33**, 2235–2243 (2012)
8. Polprasert, C.: Organic Waste

9. Bernal, M.P., Sommer, S.G., Chadwick, D., Qing, C., Guoxue, L., Michel, F.C.: Chapter three - current approaches and future trends in compost quality criteria for agronomic, environmental, and human health benefits. In: Sparks, D.L. (ed.) Advances in Agronomy, vol. 144, pp. 143–233. Academic Press (2017)

10. Chen, R., Wang, Y., Wang, W., Wei, S., Jing, Z., Lin, X.: N2O emissions and nitrogen transformation during windrow composting of dairy manure. J. Environ. Manage. **160**, 121–127 (2015)

11. DeLaune, P.B., Moore, P.A., Daniel, T.C., Lemunyon, J.L.: Effect of chemical and microbial amendments on ammonia volatilization from composting poultry litter. J. Environ. Qual. **33** (2), 728–734 (2004)

12. Kalemelawa, F., et al.: An evaluation of aerobic and anaerobic composting of banana peels treated with different inoculums for soil nutrient replenishment. Bioresour. Technol. **126**, 375–382 (2012)

13. Parkinson, R., Gibbs, P., Burchett, S., Misselbrook, T.: Effect of turning regime and seasonal weather conditions on nitrogen and phosphorus losses during aerobic composting of cattle manure. Bioresour. Technol. **91**(2), 171–178 (2004)

14. Awasthi, M.K., Pandey, A.K., Khan, J., Bundela, P.S., Wong, J.W.C., Selvam, A.: Evaluation of thermophilic fungal consortium for organic municipal solid waste composting. Bioresour. Technol. **168**, 214–221 (2014)

15. Kalamdhad, A.S., Kazmi, A.A.: Effects of turning frequency on compost stability and some chemical characteristics in a rotary drum composter. Chemosphere **74**(10), 1327–1334 (2009)

16. Ogunwande, G.A., Osunade, J.A., Adekalu, K.O., Ogunjimi, L.A.O.: Nitrogen loss in chicken litter compost as affected by carbon to nitrogen ratio and turning frequency. Bioresour. Technol. **99**(16), 7495–7503 (2008)

17. Getahun, T., Nigusie, A., Entele, T., Gerven, T.V., der Bruggen, B.V.: Effect of turning frequencies on composting biodegradable municipal solid waste quality. Resour. Conserv. Recycl. **65**, 79–84 (2012)

18. Zhao, X., Li, B., Ni, J., Xie, D.: Effect of four crop straws on transformation of organic matter during sewage sludge composting. J. Integr. Agric. **15**(1), 232–240 (2016)

19. Raut, M.P., William, S.P., Bhattacharyya, J.K., Chakrabarti, T., Devotta, S.: Microbial dynamics and enzyme activities during rapid composting of municipal solid waste–a compost maturity analysis perspective. Bioresour. Technol. **99**(14), 6512–6519 (2008)

20. Kulikowska, D.: Kinetics of organic matter removal and humification progress during sewage sludge composting. Waste Manag. **49**, 196–203 (2016)

21. Turan, N.G., Akdemir, A., Ergun, O.N.: Emission of volatile organic compounds during composting of poultry litter. Water Air Soil Pollut. **184**(1–4), 177–182 (2007)

22. Chen, L., De Haro, M., Moore, A., Falen, C.: The Composting Process: Dairy Compost Production and Use in Idaho CIS 1179. Univ. Ida. (2011)

23. Pace, M.G., Miller, B.E., Farrell-Poe, K.L.: The composting process (1995)

24. Iqbal, M.K., Nadeem, A., Sherazi, F., Khan, R.A.: Optimization of process parameters for kitchen waste composting by response surface methodology. Int. J. Environ. Sci. Technol. **12**(5), 1759–1768 (2015)

25. Igoni, A.H., Ayotamuno, M.J., Eze, C.L., Ogaji, S.O.T., Probert, S.D.: Designs of anaerobic digesters for producing biogas from municipal solid-waste. Appl. Energy **85**(6), 430–438 (2008)

26. Johnson, G.A., Qian, Y.L., Davis, J.G.: Effects of compost topdressing on turf quality and growth of Kentucky bluegrass. Appl. Turfgrass Sci. **3**(1) (2006)

27. Lazcano, C., Gómez-Brandón, M., Domínguez, J.: Comparison of the effectiveness of composting and vermicomposting for the biological stabilization of cattle manure. Chemosphere **72**(7), 1013–1019 (2008)
28. Chan, M.T., Selvam, A., Wong, J.W.C.: Reducing nitrogen loss and salinity during 'struvite' food waste composting by zeolite amendment. Bioresour. Technol. **200**, 838–844 (2016)
29. Zhang, L., Sun, X.: Changes in physical, chemical, and microbiological properties during the two-stage co-composting of green waste with spent mushroom compost and biochar. Bioresour. Technol. **171**, 274–284 (2014)
30. Nakasaki, K., Tran, L.T.H., Idemoto, Y., Abe, M., Rollon, A.P.: Comparison of organic matter degradation and microbial community during thermophilic composting of two different types of anaerobic sludge. Bioresour. Technol. **100**(2), 676–682 (2009)
31. Petric, I., Helić, A., Avdić, E.A.: Evolution of process parameters and determination of kinetics for co-composting of organic fraction of municipal solid waste with poultry manure. Bioresour. Technol. **117**, 107–116 (2012)
32. Chen, C.-Y., Mei, H.-C., Cheng, C.-Y., Lin, J.-H., Chung, Y.-C.: Enhancing the conversion of organic waste into biofertilizer with thermophilic bacteria. Environ. Eng. Sci. **29**(7), 726–730 (2012)
33. Microbial conversion of food wastes for biofertilizer production with thermophilic lipolytic microbes - ScienceDirect. https://www.sciencedirect.com/science/article/pii/S0960148106001030. Accessed 03 Apr 2018
34. Nakasaki, K., Uehara, N., Kataoka, M., Kubota, H.: The use of Bacillus licheniformis HA1 to accelerate composting of organic wastes. Compost Sci. Util. **4**(4), 47–51 (1996)
35. Sundberg, C., Smårs, S., Jönsson, H.: Low pH as an inhibiting factor in the transition from mesophilic to thermophilic phase in composting. Bioresour. Technol. **95**(2), 145–150 (2004)
36. Tang, J.-C., Kanamori, T., Inoue, Y., Yasuta, T., Yoshida, S., Katayama, A.: Changes in the microbial community structure during thermophilic composting of manure as detected by the quinone profile method. Process Biochem. **39**(12), 1999–2006 (2004)
37. Earthworm–microorganism interactions: a strategy to stabilize domestic wastewater sludge - ScienceDirect. https://www.sciencedirect.com/science/article/pii/S0043135410000217. Accessed 03 Apr 2018
38. Nakasaki, K., Hirai, H.: Temperature control strategy to enhance the activity of yeast inoculated into compost raw material for accelerated composting. Waste Manag **65**, 29–36 (2017)
39. Xiao, Y., et al.: Continuous thermophilic composting (CTC) for rapid biodegradation and maturation of organic municipal solid waste. Bioresour. Technol. **100**(20), 4807–4813 (2009)
40. Ramachandra, T.V., Bharath, H.A., Kulkarni, G., Han, S.S.: Municipal solid waste: generation, composition and GHG emissions in Bangalore, India. Renew. Sustain. Energy Rev. **82**, 1122–1136 (2018)
41. Smårs, S., Gustafsson, L., Beck-Friis, B., Jönsson, H.: Improvement of the composting time for household waste during an initial low pH phase by mesophilic temperature control. Bioresour. Technol. **84**(3), 237–241 (2002)
42. Nakasaki, K., Yaguchi, H., Sasaki, Y., Kubota, H.: Effects of pH control on composting of garbage. Waste Manag. Res. **11**(2), 117–125 (1993)
43. Hagemann, N., et al.: Effect of biochar amendment on compost organic matter composition following aerobic compositing of manure. Sci. Total Environ. **613**, 20–29 (2018)
44. Sanchez-Monedero, M.A., Cayuela, M.L., Roig, A., Jindo, K., Mondini, C., Bolan, N.: Role of biochar as an additive in organic waste composting. Bioresour. Technol. **247**, 1155–1164 (2018)

45. Waqas, M., Nizami, A.S., Aburiazaiza, A.S., Barakat, M.A., Ismail, I.M., Rashid, M.I.: Optimization of food waste compost with the use of biochar. J. Environ. Manage. (2017)
46. Zhang, L., Sun, X.: Addition of seaweed and bentonite accelerates the two-stage composting of green waste. Bioresour. Technol. **243**, 154–162 (2017)
47. Clarke, J.: Green Waste. pdf (2018)
48. Zhang, L., Sun, X.: Effects of bean dregs and crab shell powder additives on the composting of green waste. Bioresour. Technol.
49. Bustamante, M.A., Ceglie, F.G., Aly, A., Mihreteab, H.T., Ciaccia, C., Tittarelli, F.: Phosphorus availability from rock phosphate: combined effect of green waste composting and sulfur addition. J. Environ. Manage. **182**, 557–563 (2016)
50. Belyaeva, O.N., Haynes, R.J., Sturm, E.C.: Chemical, physical and microbial properties and microbial diversity in manufactured soils produced from co-composting green waste and biosolids. Waste Manag **32**(12), 2248–2257 (2012)
51. Wang, Q., et al.: Improvement of pig manure compost lignocellulose degradation, organic matter humification and compost quality with medical stone. Bioresour. Technol. **243**, 771–777 (2017)
52. Zhang, L., Sun, X.: Addition of fish pond sediment and rock phosphate enhances the composting of green waste. Bioresour. Technol. **233**, 116–126 (2017)
53. Zhang, L., Sun, X.: Influence of bulking agents on physical, chemical, and microbiological properties during the two-stage composting of green waste. Waste Manag **48**, 115–126 (2016)
54. Godlewska, P., Schmidt, H.P., Ok, Y.S., Oleszczuk, P.: Biochar for composting improvement and contaminants reduction. A review. Bioresour. Technol. **246**, 193–202 (2017)
55. Xiao, R., et al.: Recent developments in biochar utilization as an additive in organic solid waste composting: a review. Bioresour. Technol. **246**, 203–213 (2017)
56. Additives aided composting of green waste: Effects on organic matter degradation, compost maturity, and quality of the finished compost - PDF Free Download. tiptiktak.com. https://tiptiktak.com/additives-aided-composting-of-green-waste-effects-on-organic-matter-degradation.html. Accessed 19 June 2018
57. Nakasaki, K., Araya, S., Mimoto, H.: Inoculation of Pichia kudriavzevii RB1 degrades the organic acids present in raw compost material and accelerates composting. Bioresour. Technol. **144**, 521–528 (2013)

Optimization Algorithms Applied to Anaerobic Digestion Process of Olive Mill Wastewater

Literature Review

Echarrafi Khadija[(✉)], Zouitina Manale, El Hassani Ibtisam,
and El Hajji Mounia

ENSEM, Hassan II University, Casablanca, Morocco
k.echarrafi@ensem.ac.ma

Abstract. Agriculture has always been a strategic sector for socioeconomic development in Morocco. Indeed, the green Morocco plan aims to extend the area of olive trees and set up new modern crushing plants in order to increase productivity and competitiveness of the olive industry. Olive mill solid waste (OMSW) and olive mill wastewater (OMW) are two types of waste generated by this industry. These wastes are rich in organic matter, but their discharge without pretreatment in nature has a toxic effect on the natural environment (soil, air, and water) because of their high acidity due to their polyphenol content.

White Biotechnology, particularly anaerobic digestion (AD), remains an effective way that uses the effluents mentioned above as a raw material for producing a renewable energy such as biogas (CH4). The aim of this work is to present a literature review of anaerobic digestion process, its influencing parameters as well as optimization algorithms used for optimizing the process and predicting gas yield.

Keywords: OMW · Anaerobic digestion · Optimization algorithms

1 Introduction

OMW is considered as one of the most serious ecological problems because of its high content phenolic compounds (source of high acidity), suspended solids, volatile fatty acids, nitrogen compounds and biodegradation resistance [1]. But, its biogas recovery and energy potential are higher e.g. its biogas yield is estimated at 57 m^3/m^3 [2] and energy potential at 14876 toe/year [3].

This type of waste is generated from olive oil extraction plants. In Morocco, this industry contributes about 5% in the national agricultural GDP (Gross Domestic Product). Indeed, Morocco has an olive area of 784,000 producing 1,500,000 tons of olives and 160,000 tons of olive oil [4].

For example, Fez region accounts for 39% of the national olive production and contains 39 modern and traditional olive oil extraction plants. These plants generate between 80,000 m^3 and 160,000 m^3 of Olive Mill Wastewater (OMW) discharged directly into the watercourses of Oued Sebou [5]. The amount of OMW depends on the oil extraction system and the type of manufacturing unit (traditional or modern).

© Springer Nature Switzerland AG 2019
M. Ezziyyani (Ed.): AI2SD 2018, AISC 913, pp. 84–94, 2019.
https://doi.org/10.1007/978-3-030-11881-5_8

Indeed, it was estimated at 40 kg of OMW/100 kg of olives for traditional unit versus 150 kg/100 kg for industrial unit [6].

In order to reduce the impact of these influents on the environment and convert a non-added value matter into a renewable energy, several biotechnology and bioprocesses engineering can be used for this purpose. AD is one of these bioprocesses discussed in this study.

The choice of AD as a suitable bioprocess engineering to treat these effluents is justified by strategic orientations of Morocco such as the tripartite pact signed between the Ministry of Agriculture, the Ministry of Economy and Finance and the Ministry Delegate for the Environment in 2015 which aimed to reduce the impact of OMW and Olive Mill Solid Waste (OMSW) to 80% by 2020 (730 000 for OMSW and 900 000 for OMW) [7] and the Moroccan orientations to produce 1,400 MW of renewable energy by 2020.

Converting OMW into biogas by using AD process is the ultimate goal, but improving performance and parameters monitoring of the process are also an important engineering task. Smart tools are increasingly used in the design, modeling and control of complex systems such as bioprocesses. These tools belong to soft computing techniques namely: Neural Networks Algorithms, Fuzzy Logic Model and Genetic Algorithms. This paper shows the algorithms used in the literature in order to optimize AD parameters and predict methane yield.

2 Anaerobic Digestion Process

AD process is a biological process converting the organic matter into a biogas and digestate it by involving a group of microorganism in the absence of oxygen [8]. It is divided into four stages: hydrolysis, acedogenesis, acetogenesis, and methanogenesis, each of them is carried out by a specific bacterial consortium as shown in Table 1. Furthermore, it generates a biogas composed mainly of methane (55 to 70%) and carbon dioxide (45 to 30%). The resulting biogas is distinguished from the natural one by its high CO2 content and the presence of other chemical compounds such as water vapor (H_2O), ammonia (NH_3) and hydrogen sulphide (H_2S) [9].

Table 1. Anaerobic digestion process stages

Process steps	Type of bacteria involved	Outputs
Hydrolysis	Hydrolysis bacteria	Monosaccharide, amino and fatty acids
Acidogenesis (increased acidity)	Acidic bacteria	Organic acids and carbon dixiode
Acetogenesis (acetic acid formation)	Acetic bacteria	Acetic acid, carbon dixide and hydrogen
Methanogenesis (methane formation)	Methanogenic bacteria	Methane

3 Oil Mill Wastewater

OMW is the aqueous residual liquid from olive oil extraction. This effluent is relatively rich in organic matter but it creates a real problem for the olive industry due to their toxicity caused by phenolic compounds. It has a negative effect on natural landscape caused by their dark red color to the black one. The shade of color depends on the type of olive extraction system and degradation state [11]. It contains a variety of organic fractions such as sugars, tannins, polyphenols, polyalcohols, and pectins [10].

OMW is distinguished from other types of waste by its high phenolic compound content. These compounds have an inhibitory effect on COD degradation (Chemical Oxygen Demand) especially in methanogenic stage where only 20–30% of phenolic compounds are degraded [12]. Besides, Syringic acid, p-hydroxyphenylacetic acid, vanillic acid, veratric acid, caffeic acid, protocathechuic acid, p-coumaric acid and cinnamic acid are considered as the most phenolic acids in OMW content [13].

AD is carried out by three types of bacteria such as: hydrolytic, acetogenic, and methanogenic bacteria, each of them is sensitive to one or more phenolic compounds from a determined concentration e.g. at 40.25 g/l of caffeic acid concentration, acetogenic bacteria become sensitive and at 0.12 g/l of caffeic, ferulic and cumaric acids concentrations, methanogenic bacteria become sensitive [14].

Before introducing OMW as a substrate in AD, it should be pre-treated using one or a mixture of pretreatment methods in order to reduce its phenolic acids concentrations e.g. the AD of OMW coupling with an aerobic pretreatment method using P.ostreatus or white rot fungi can remove from 50% to 78.3% of phenolic compounds concentrations [15, 16].

Finally, AD of OMW is considered as a suitable process for many reasons: producing renewable energy, reducing the sludge production, producing high energy recovery as well as reducing the carbon footprint as shown in Table 2.

Table 2. Organic waste potential and its energy equivalent in Morocco [6]

	Theoretical potential in t/year		Available potential in t/year	Available potential in tep[a]	Carbon footprint (t/year)
Agri-food industry waste	Dairy industry	400000	8000	412,8	1888
	Olive industry	505800	493155	25446	116384,6
	Slaughterhouses	3810633	3810633	196632,8	899309,4

[a]The calculations are based on a theoretical average yield of 100 m^3/t and methane content of 60% and 1.tep = 11628 kWh

4 Parameters Affecting Anaerobic Digestion

4.1 Temperature

There are three favorable temperature ranges for the growth of anaerobic bacteria: Psychrophilic (<20 °C), Mesophilic (>25 °C) and Thermophilic (50–60 °C).

The performance and stability of AD process depend on temperature range which has an influence on the bacteria growth rate and activity [17–19]. Indeed, higher temperatures have a significant effect on the digestion process more than the lower ones [20]. Also, the amount of energy provided by thermophilic digestion is higher than the mesophilic one as shown in Table 3 [21]. However, the cost of the process is estimated at 7.8 cents/1,000 ft^3 at 60 °C versus 11.0 cents at 35 °C [22]. Then, the biodegradability rate is higher in thermophilic conditions than in the mesophilic ones such as it decreases, respectively, from 92.5 to 84.7% and from 94.6 to 90.4% in HRT (Hydraulic Retention Time) of 25 days [23]. The methane yield coefficient is about 0.29 and 0.23 L, respectively, for the thermophilic and mesophilic reactors [23]. As a result, thermophilic conditions are suitable to accelerate digestion rate and enhance methane yield.

Table 3. Comparison of CH4 yield between thermophilic and mesophilic conditions

Loading rate (g VS l−1 reactor volume per day)	Gas composition (%)				Gas production (1 day^{-1})					
	Mesophilic		Thermophilic		Total		Mesophilic		Thermophilic	
	CH$_4$	CO$_2$	CH$_4$	CO$_2$	Mesophilic	Thermophilic	CH$_4$	CO$_2$	CH$_4$	CO$_2$
3	58,9	38	85	39,1	3,18	3,6	1,87	1,21	2,09	1,41
6	58,7	38,5	58,1	39	5,58	6,67	3,27	2,15	3,87	2,6
9	54,9	41,8	58,9	37,2	6,45	9,74	3,54	2,7	5,73	3,62
12	52,8	43,8	60,2	36,4	7,22	11,68	3,81	3,16	7,03	4,25

4.2 Volatile Fatty Acid (VFA) Concentration

VFAs are among the most important parameters in the monitoring of the AD process. There are two types of VFAs contributing to methane production; butyric acid and acetic acid which has 70% as contribution ratio [24]. The inhibition of methanogenesis can be caused by high VFA concentration because methanogenic bacteria are vulnerable to propionic acid when its concentration reaches up the range of 1,000–2,000 mg/L [25–27]. Research studies found that at 33 g/L of VFA concentrations, hydrolysis of bio waste can be inhibited.

The accumulation of VFAs can lead to a decrease in pH value which can decelerate the hydrolysis and acidogenesis phase [28]. Some studies show that Valerate accumulation is among of the most toxic VFAs and the functioning of the reactor can be affected by propionate, butyrate and acetate accumulations [29]. According to the literature, the performance and stability of the reactor are ensured by remaining VFA concentrations at their optimal values.

4.3 pH

pH is considered as one of the most important and sensitive parameters in AD process control. Adrie Veeken, Sergey Kalyuzhnyi, Heijo Scharff, and Bert Hamelers [30] proved that the inhibitory effect of this parameter in the hydrolysis stage when it's lower than (5.0–5.5) range. Figure 1 describes the positive linear relationship between hydrolysis rate and pH parameter and it shows that the hydrolysis activity start at pH 5.

However, pH has a significant effect of pH on lipid degradation [27]. Indeed at pH 8.5, lipids are completely degraded whereas not degraded at all at pH 6 [27]. Finally, according to the AD literature pH 7 is considered as an optimal value for the methanogenesis phase [31].

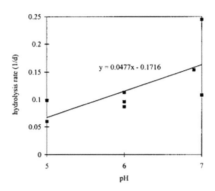

Fig. 1. Hydrolysis Rate as Function of pH [30]

4.4   Organic Loading Rate (Dilution Rate) and Hydraulic Retention Time

HRT is the average period of time during which a portion of the substance remains in the digester. It's a primordial parameter as influences digester efficiency. According to the AD literature, this parameter fluctuates between 10 and 60 days, but an HRT of 20 to 25 days is applied to closely controlled anaerobic digester because a short HRT may cause washout phenomena which affect bacteria activity and stability. This parameter is correlated with operating digestion parameters especially COD removal [32].

Otherwise, OLR is the amount of Volatile Solids (VS) loaded per time period (usually daily) per volume of digester (ft^3). This process variable depends on the type of waste introduced in the digester. In addition, almost all reactors design have been made to be able to contain and treat from 0.1 to 0.4 lb of VS/ft^3 [33].

OLR and HRT are inversely proportional because the increasing of OLR requires the lowering of HRT even though low HRT slows down bioreactor recovery [34] and a low OLR leads to a good COD reduction [14]. One of research studies suggested an OLR of 1.4 kg $VS/(m^3.d)$ as the optimal value which can produce 0.25 m^3 CH4/kg VS input of methane production rate [35].

4.5   Rapport C/N

The C/N ratio is an indicator of the ability of an organic matter to decompose. The initial characterization of organic matter plays a crucial role in order to determine its nitrogen and Carbone content. The determining of C/N rate avoids the problem of the AD process limitation caused by imbalanced nutrients. The authors consider that the optimal value of C/N ratio is between 20–30 [33, 36] or 22–33 [37, 38].

The C/N ratio has two undesirable effects: if it's higher it affects growth and reproduction of anaerobic bacteria due to nutrient deficiency, and if it's too lower, the inhibitor effect of ammonia (total or free) on methanogenesis takes place [39].

4.6   Moisture

Moisture plays a crucial role in the AD process. Indeed its high ratio makes the digestion more fluid, but it's not easy to ensure a constant indicator of water's amount throughout the digestion cycle [40]. The authors proved that methane production is good at 90–96% of moisture in the AD of sludge [39], but when moisture content of sludge is less than 91.1%, the methane production is decreased [41]. Another study shows that in the range of 60–80% of moisture content, the highest methane production may be generated [42]. The anaerobic co-digestion between 94.8% of OMW moisture content and 94.3% of Cheese whey leads to a methane production included between 68–75% [14].

5   Wastewater Anaerobic Digestion Optimization Methods

Soft computing tools are widely used in bioprocess optimization. This section presents some intelligent tools applied in the optimization of AD process because this one is characterized by non-linear behavior of its inputs and outputs parameters. According to the literature, a lot of algorithms are applied to AD process in order to monitor its biological and physical-chemical behavior. Among these tools, we find ANN (Artificial Neurons Network) [43], ANN combined with PCA (Principal Component Analysis) [44] or GA (Genetic Algorithms) [45], ADM1 (Anaerobic Digestion Model No 1) [46], and Fuzzy Logic Model [47].

The application of the above mentioned tools are widely applied to AD process. Indeed, Qdais, Hani, and Shatnawi [48] have used GA (Genetic Algorithms) with multi-layer ANN pattern with two hidden layers and sigmoid activation function in order to predict methane yield. Indeed, they founded that the optimal amount of methane yield is around 77% with an adjustment of 0.87 as a correlation coefficient. This yield is obtained at the following input parameters 36 °C, TS 6.6%, TVS 52.8% and pH 6.4.

In addition, GA-ANN is also used by Kana, Oloke, Lateef, and Adesiyan [49] for optimizing biogas production generated from co-digestion of a set of solid wastes. Indeed, they got a rate of 8.64% as an enhancement of non-optimized process.

Another study has applied ANN algorithms combined with other pattern such as PCA in order to reduce inputs vector of the process and predict CH4 and COD concentrations optimum values Fig. 2 [50].

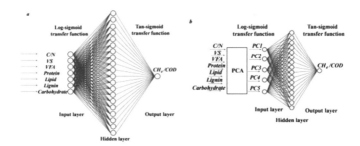

Fig. 2. ANN and PC-ANN architecture for predicting CH4 yield and COD concentrations

Kaan Yetilmezsoy and Zehra Sapci-Zengin have applied the ANN algorithms coupled with PCA on textile wastewater in order to predict COD Removal Efficiency (CODRE) of UASB (Upflow Anaerobic Sludge Blanket) reactors. They use 9 inputs and "purelin" as transfer function as shown in Fig. 3.

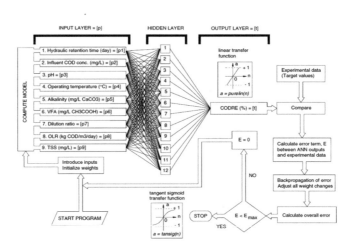

Fig. 3. Architecture of ANN coupled with BP (back-propagation algorithm) for the prediction of CODRE

Otherwise, JOO-HWA TAY and XIYUE ZHANG [51] have used neural fuzzy model (Fig. 4) in laboratory scale on three types of high-rate anaerobic reactors such as Anaerobic Fluidized Bed Reactor (AFBR), Anaerobic Filter (AF), and UASB. They use as inputs system the following parameters: Organic Loading Rate (OLR), Hydraulic Loading Rate (HLR) and Alkalinity Loading Rate (ALR); Volumetric Methane Production (VMP), effluent TOC (Total Organic Carbon) and total VFA concentration. They have obtained good predicting results in terms of correlation coefficients (R^2), e.g. they founded, respectively, 0.987, 0.975 and 0.975 for VMP, TOC and VFA using AF system.

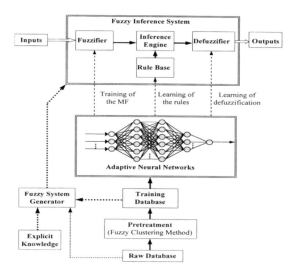

Fig. 4. Concept of neural fuzzy model concept applied in anaerobic wastewater treatment systems

The ADM1 [46] model has been widely used in several researches of AD process optimization for predicting the behavior of its inputs or outputs parameters. The disadvantage of this model is that it gives a large gap between the predicted values and the experimental ones. That means it overestimates or underestimates the AD parameters [52]. Unlike some modifications of this model can provide satisfactory results [53].

There are many optimization algorithms belonging to the artificial intelligence such as Colony Optimization (ACO) [56] and Particle Swarm Optimization (PSO) [57]. All these algorithms provide satisfactory results because they explore all interval solutions and give the best one.

6 Conclusion

According to the literature, AD process is one of the most efficient processes for managing waste with high moisture content, wastewater or solid waste, and producing renewable energy. An optimization of the AD process parameters is required for ensuring the efficiency of the process (producing a high methane yield with low cost). This optimization can be done using one of the optimization tools mentioned previously.

As the OMW is an environmental issue especially during the olive harvest season and its treatment by an optimized AD process is not well evocated in the literature, the AD of OMW optimization using the artificial intelligence algorithms will be our future work.

References

1. Dareioti, M.A., Dokianakis, S.N., Stamatelatou, K., Zafiri, C., Kornaros, M.: Exploitation of olive mill wastewater and liquid cow manure for biogas production. Waste Manag. **30**(10), 1841–1848 (2010)
2. Etude sur les potentiels de biomasse pour la région Souss-Massa-Drâa et la province d'Essaouira (2010)
3. Elamin, P.A.M.: Potentiel des déchets organiques et valorisation énergétique au Maroc
4. Ministère de l'agriculture, de la pêche maritime, du développement rural et des eaux et des forêts. http://www.agriculture.gov.ma/pages/acces-fillieres/filiere-oleicole
5. Idrissi, M.: Trituration des olives : comment arrêter la pollution par les margines (2016). http://www.leseco.ma/415-regions/41629-trituration-des-olives-comment-arreter-la-pollution-par-les-margines.html
6. Benyahia, N., Zein, K.: Analyse des problèmes de l'industrie de l'huile d'olive et solutions récemment développées, Sustain. Bus. Assoc. Lausanne, pp. 1–8 (2003)
7. Examen national de l'export vert du Maroc : produits oléicoles, romarin et thym (2017)
8. Moletta, R.: La méthanisation. Tec et Doc (2015)
9. Béline, F., Girault, R., Peu, P., Trémier, A., Téglia, C., Dabert, P.: Enjeux et perspectives pour le développement de la méthanisation agricole en France. Sci. Eaux Territ. **2**, 34–43 (2012)
10. Robles, A., Lucas, R., de Cienfuegos, G.A., Gálvez, A.: Biomass production and detoxification of wastewaters from the olive oil industry by strains of Penicillium isolated from wastewater disposal ponds. Bioresour. Technol. **74**(3), 217–221 (2000)
11. Hamdi, M.: Future prospects and constraints of olive mill wastewaters use and treatment: a review. Bioprocess. Eng. **8**(5–6), 209–214 (1993)
12. Beccari, M., Majone, M., Torrisi, L.: Two-reactor system with partial phase separation for anaerobic treatment of olive oil mill effluents. Water Sci. Technol. **38**(4–5), 53–60 (1998)
13. Lesage-Meessen, L., et al.: Simple phenolic content in olive oil residues as a function of extraction systems. Food Chem. **75**(4), 501–507 (2001)
14. Martinez-Garcia, G., Johnson, A.C., Bachmann, R.T., Williams, C.J., Burgoyne, A., Edyvean, R.G.J.: Two-stage biological treatment of olive mill wastewater with whey as co-substrate. Int. Biodeterior. Biodegradation **59**(4), 273–282 (2007)
15. Blika, P.S., et al.: Pretreatment of olive mill wastewater in a fungal trickling filter In: Proceedings of the Third International Conference on Water Resources in Mediterranean Basin, Tripoli, Lebanon (2006)
16. Fountoulakis, M.S., Dokianakis, S.N., Kornaros, M.E., Aggelis, G.G., Lyberatos, G.: Removal of phenolics in olive mill wastewaters using the white-rot fungus Pleurotus ostreatus. Water Res. **36**(19), 4735–4744 (2002)
17. Lettinga, G., et al.: High-rate anaerobic waste-water treatment using the UASB reactor under a wide range of temperature conditions. Biotechnol. Genet. Eng. Rev. **2**(1), 253–284 (1984)
18. Ziganshin, A.M., Liebetrau, J., Pröter, J., Kleinsteuber, S.: Microbial community structure and dynamics during anaerobic digestion of various agricultural waste materials. Appl. Microbiol. Biotechnol. **97**(11), 5161–5174 (2013)
19. Labatut, R.A., Angenent, L.T., Scott, N.R.: Conventional mesophilic vs. thermophilic anaerobic digestion: a trade-off between performance and stability? Water Res. **53**, 249–258 (2014)
20. Labatut, R.A., Gooch, C.A.: Monitoring of anaerobic digestion process to optimize performance and prevent system failure (2014)

21. Mackie, R.I., Bryant, M.P.: Anaerobic digestion of cattle waste at mesophilic and thermophilic temperatures. Appl. Microbiol. Biotechnol. **43**(2), 346–350 (1995)
22. Varel, V.H., Isaacson, H.R., Bryant, M.P.: Thermophilic methane production from cattle waste. Appl. Environ. Microbiol. **33**(2), 298–307 (1977)
23. Borja, R., Martin, A., Banks, C.J., Alonso, V., Chica, A.: A kinetic study of anaerobic digestion of olive mill wastewater at mesophilic and thermophilic temperatures. Environ. Pollut. **88**(1), 13–18 (1995)
24. Wijekoon, K.C., Visvanathan, C., Abeynayaka, A.: Effect of organic loading rate on VFA production, organic matter removal and microbial activity of a two-stage thermophilic anaerobic membrane bioreactor. Bioresour. Technol. **102**(9), 5353–5360 (2011)
25. Lee, D.-J., et al.: Effect of volatile fatty acid concentration on anaerobic degradation rate from field anaerobic digestion facilities treating food waste leachate in South Korea. J. Chem. **2015**, 9 (2015)
26. Siegert, I., Banks, C.: The effect of volatile fatty acid additions on the anaerobic digestion of cellulose and glucose in batch reactors. Process Biochem. **40**(11), 3412–3418 (2005)
27. Beccari, M., Bonemazzi, F., Majone, M., Riccardi, C.: Interaction between acidogenesis and methanogenesis in the anaerobic treatment of olive oil mill effluents. Water Res. **30**(1), 183–189 (1996)
28. Wong, Y.-S., Teng, T.T., Ong, S.-A., Norhashimah, M., Rafatullah, M., Lee, H.-C.: Anaerobic acidogenesis biodegradation of palm oil mill effluent using suspended closed anaerobic bioreactor (SCABR) at mesophilic temperature. Procedia Environ. Sci. **18**, 433–441 (2013)
29. Mechichi, T., Sayadi, S.: Evaluating process imbalance of anaerobic digestion of olive mill wastewaters. Process Biochem. **40**(1), 139–145 (2005)
30. Veeken, A., Kalyuzhnyi, S., Scharff, H., Hamelers, B.: Effect of pH and VFA on hydrolysis of organic solid waste. J. Environ. Eng. **126**(12), 1076–1081 (2000)
31. Ağdağ, O.N., Sponza, D.T.: Co-digestion of industrial sludge with municipal solid wastes in anaerobic simulated landfilling reactors. Process Biochem. **40**(5), 1871–1879 (2005)
32. Zhang, S.J., Liu, N.R., Zhang, C.X.: Study on the performance of modified UASB process treating sewage. Adv. Mater. Res. **610**, 2174–2178 (2013)
33. Mattocks, R.: Understanding biogas generation (1984)
34. Trnovec, W., Britz, T.J.: Influence of organic loading rate and hydraulic retention time on the efficiency of a UASB bioreactor treating a canning factory effluent. Water S. A. **24**(2), 147–152 (1998)
35. Babaee, A., Shayegan, J.: Effect of organic loading rates (OLR) on production of methane from anaerobic digestion of vegetables waste. In: World Renewable Energy Congress-Sweden, 8–13 May 2011, Linköping Sweden, no. 57, pp. 411–417 (2011)
36. Li, Y., Park, S.Y., Zhu, J.: Solid-state anaerobic digestion for methane production from organic waste. Renew. Sustain. Energy Rev. **15**(1), 821–826 (2011)
37. Yen, H.-W., Brune, D.E.: Anaerobic co-digestion of algal sludge and waste paper to produce methane. Bioresour. Technol. **98**(1), 130–134 (2007)
38. Habiba, L., Hassib, B., Moktar, H.: Improvement of activated sludge stabilisation and filterability during anaerobic digestion by fruit and vegetable waste addition. Bioresour. Technol. **100**(4), 1555–1560 (2009)
39. Lin, J., et al.: Effects of mixture ratio on anaerobic co-digestion with fruit and vegetable waste and food waste of China. J. Environ. Sci. **23**(8), 1403–1408 (2011)
40. Hernández-Berriel, M.C., Márquez-Benavides, L., González-Pérez, D.J., Buenrostro-Delgado, O.: The effect of moisture regimes on the anaerobic degradation of municipal solid waste from Metepec (Mexico). Waste Manag. **28**, S14–S20 (2008)

41. Fujishima, S., Miyahara, T., Noike, T.: Effect of moisture content on anaerobic digestion of dewatered sludge: ammonia inhibition to carbohydrate removal and methane production. Water Sci. Technol. **41**(3), 119–127 (2000)
42. Munasinghe, R.: Effect of hydraulic retention time on landfill leachate and gas characteristics. University of British Columbia (1997)
43. Agatonovic-Kustrin, S., Beresford, R.: Basic concepts of artificial neural network (ANN) modeling and its application in pharmaceutical research. J. Pharm. Biomed. Anal. **22**(5), 717–727 (2000)
44. Bro, R., Smilde, A.K.: Principal component analysis. Anal. Methods **6**(9), 2812–2831 (2014)
45. Gen, M., Cheng, R., Lin, L.: Network models and optimization: multiobjective genetic algorithm approach. Springer Science & Business Media (2008)
46. Batstone, D.J., et al.: The IWA anaerobic digestion model no 1 (ADM1). Water Sci. Technol. **45**(10), 65–73 (2002)
47. Akkurt, S., Tayfur, G., Can, S.: Fuzzy logic model for the prediction of cement compressive strength. Cem. Concr. Res. **34**(8), 1429–1433 (2004)
48. Qdais, H.A., Hani, K.B., Shatnawi, N.: Modeling and optimization of biogas production from a waste digester using artificial neural network and genetic algorithm. Resour. Conserv. Recycl. **54**(6), 359–363 (2010)
49. Kana, E.B.G., Oloke, J.K., Lateef, A., Adesiyan, M.O.: Modeling and optimization of biogas production on saw dust and other co-substrates using artificial neural network and genetic algorithm. Renew. Energy **46**, 276–281 (2012)
50. Li, H., et al.: Estimating the fates of C and N in various anaerobic codigestions of manure and lignocellulosic biomass based on artificial neural networks. Energy Fuels **30**(11), 9490–9501 (2016)
51. Tay, J.-H., Zhang, X.: A fast predicting neural fuzzy model for high-rate anaerobic wastewater treatment systems. Water Res. **34**(11), 2849–2860 (2000)
52. Parker, W.J.: Application of the ADM1 model to advanced anaerobic digestion. Bioresour. Technol. **96**(16), 1832–1842 (2005)
53. Boubaker, F., Ridha, B.C.: Modelling of the mesophilic anaerobic co-digestion of olive mill wastewater with olive mill solid waste using anaerobic digestion model No. 1 (ADM1). Bioresour. Technol. **99**(14), 6565–6577 (2008)
54. Turkdogan-Aydınol, F.I., Yetilmezsoy, K.: A fuzzy-logic-based model to predict biogas and methane production rates in a pilot-scale mesophilic UASB reactor treating molasses wastewater. J. Hazard. Mater. **182**(1–3), 460–471 (2010)
55. Puñal, A., Palazzotto, L., Bouvier, J.C., Conte, T., Steyer, J.P.: Automatic control of volatile fatty acids in anaerobic digestion using a fuzzy logic based approach. Water Sci. Technol. **48**(6), 103–110 (2003)
56. Rao, K.R., Srinivasan, T., Venkateswarlu, C.: Mathematical and kinetic modeling of biofilm reactor based on ant colony optimization. Process Biochem. **45**(6), 961–972 (2010)
57. Garlapati, V.K., Banerjee, R.: Evolutionary and swarm intelligence-based approaches for optimization of lipase extraction from fermented broth. Eng. Life Sci. **10**(3), 265–273 (2010)

Emission Inventory in Urban Road: Case Study Tangier City

Marwane Benhadou[(✉)], Nadia Bufardi, and Abdelouahid Lyhyaouyi

Laboratory of Innovative Technologies, National School of Applied Sciences,
UAE, Tangier, Morocco
mrwbenhaddou@gmail.com, nadiabufardi@gmail.com,
lyhyaoui@gmail.com

Abstract. The impact of transport not only affects the environment but also the health of citizens. Every day our lungs filter 15 kilos of air, and if we live in a big city or next to a highway, that air will contain pollutants emitted by vehicles, which indirectly affect our health. Currently more than 50% of the world population live in cities and the main responsible for the loss of air quality are the moving sources, in the interval between 75% and 80% of the total pollution. It will be interesting to quantify this problem by making an environmental inventory to see how urban transport affects air quality. In this paper we develop a methodology to estimate traffic flows and the main goal is to create an inventory emission, in the center of city where most pedestrians and vehicles are concentrated.

Keywords: Road emission inventory · Urban areas · Dijkstra algorithm · Wardrop principle · Mobility in Tangier (Morocco)

1 Problems Identification

1.1 Introduction

Climate change refers to a change in the state of the climate that can be identified by variations in the mean and/or the variability of its properties. It persists for an extended period, typically decades or longer and it refers to any change in climate over time, whether due to natural variability or as a result of human activity [1].

A wide range of methods has been developed to estimate emissions of carbon dioxide due to urban transport [2] and the energy consumption estimation models as, instantaneous fuel consumption models, modal fuel consumption models and fuel consumption models based on average speed. Researchers have used different approaches to estimate traffic emissions, as the implementation of the model refers to the prediction of CO emissions in Athens area in Greece [3] and using CAR model in the Netherlands [4], they are different methods to model traffic flow [5] Bottom-up and top-down approaches using aerial photography and GPS surveys in order to estimate the traffic flow and the quantity of pollutant emitted annually at each traffic road link [6, 7]. Also the approach for estimating daily average traffic values (AADT) [8], using different methods as neural network [9], leads to an effective calculation of emissions

© Springer Nature Switzerland AG 2019
M. Ezziyyani (Ed.): AI2SD 2018, AISC 913, pp. 95–104, 2019.
https://doi.org/10.1007/978-3-030-11881-5_9

[10], a fuel consumption based approach and vehicle kilometer travelled by each category of vehicle [11]. Others has discussed the factors that influence the calculation of emissions, distance, speed and vehicle categories [12] to elaborate emissions models as, average speed models, traffic situation emissions models, traffic variable models, cycle variable emissions models and modal emissions models [12].

Morocco ratified the Kyoto Protocol in 2002 and affirmed its willingness to contribute to the fight against climate change [13].

The Tangier-Tetouan region (North of Morocco), whose population is around 2.8 Million, with one of the largest densities of the kingdom of Morocco, 235 Hab/Km2, is vulnerable to climate change [13]. This evolution is reflected in the generation of more complex mobility, the creation of new bus lines, increased taxis and travel, and a high rate of industrialization.

We develop a methodology for preparing an inventory of carbon dioxide emissions in a region of Tangier city.

1.2   Study Area and Objectives

We chose the center of Tangier city where pedestrians are concentrated and also is the point of passage of most vehicles.

In this area, there are shopping centers (generation of travel), Spanish institute, school that generate high pedestrian and vehicular mobility during peak hours, hospitals, cafes, restaurants, administrations and also a residential area, as depicted in Fig. 1.

We observe that there is a great interaction between cars and pedestrians, which generates congestion and consequently increase the level of CO_2 emissions affecting directly the health of citizens.

Then it will be interesting to develop an environmental inventory of the region in order to calculate the CO_2 emissions based on the flow of cars in the area caused by the urban transport.

Why an inventory emission?

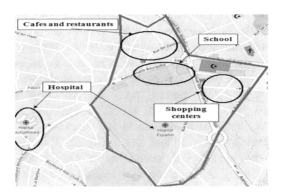

Fig. 1. The studied area

- To evaluate and monitor the impact of climate change on the territory and mitigate it.
- To anticipate regulatory constraints.
- To reduce energy dependency.
- To respond to the growing concerns of citizens.

1.3   Modeling of the Studied Area

In order to study the concerned area, we start the digitalization which means make some transformation to model the map as a directed graph, where the vertex corresponds to the intersection (in this case identified by letters) and arcs are the street sections (the arrows indicate the direction of circulation), as shown in Fig. 2. The digitalization divides the study area into zones that should be manageable for on-site work [14].

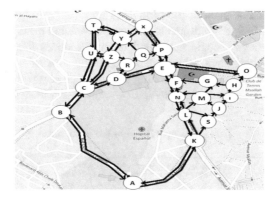

Fig. 2. Digitalization of the road

2   Methodology

To carry out this work we will develop a methodology that will provide us the carbon dioxide emitted by the vehicles circulating in this area.

2.1   Description of the Methodology

The quantity of CO2 emitted is estimated [3] according to the formula below, which multiplies the emission factor (g/km/vehicle), the volume of vehicles circulating in each segment and the distance (Km) traveled by vehicles:

$$\sum_i Vehicles_{segment\,i} * Distance_{Km} * Emission\,factors_{veh} \tag{1}$$

We represent in Fig. 3, the flow graph of the methodology followed step by step to provide us the Carbon Dioxide in the area.

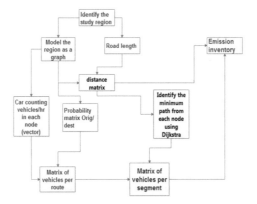

Fig. 3. The flow graph

First identify each element of the formula (1):

We calculate the length of the streets (distance traveled in meters) that constitute the graph, it's done through Google map application (the nodes that are not connected: 999999, a high number), then we calculate the hourly flow of vehicles leaving each node at peak times for a week (the counting is done manually with a manual counter), and we represent it by means of a vehicle flow vector, as shown in Table 1.

Table 1. Data related to the study area

Vector of vehicle flow in each node (vehicles/hr)											
A	B	C	D	E	F	G	H	I	J	K	L
744	984	1092	936	1224	840	180	456	516	444	1344	948
M	N	O	P	Q	R	S	T	U	X	Y	Z
276	756	1020	828	216	288	624	228	492	468	336	324

Matrix of distance (in meters)

	A	B	C	D	E	F	G	H	I	J	K	L	M	N	O	P	Q	R	S	T	U	X	Y	Z
A	0	720	999999	999999	999999	999999	999999	999999	999999	999999	433	999999	999999	999999	999999	999999	999999	999999	999999	999999	999999	999999	999999	999999
B	720	0	172	999999	999999	999999	999999	999999	999999	999999	999999	999999	999999	999999	999999	999999	999999	999999	999999	999999	999999	999999	999999	999999
C	999999	172	0	190	999999	999999	999999	999999	999999	999999	999999	999999	999999	999999	999999	999999	999999	999999	999999	999999	999999	234	999999	215
D	999999	999999	160	0	185	999999	999999	999999	999999	999999	999999	999999	999999	999999	999999	999999	125	999999	999999	999999	999999	999999	999999	999999
E	999999	999999	999999	185	0	999999	999999	999999	999999	999999	999999	999999	150	308	85	999999	999999	999999	999999	999999	999999	999999	999999	999999
F	999999	999999	999999	999999	63	0	999999	999999	999999	999999	999999	999999	999999	999999	999999	999999	999999	999999	999999	999999	999999	999999	999999	999999
G	999999	999999	999999	999999	999999	130	0	112	999999	999999	999999	999999	999999	999999	999999	999999	999999	999999	999999	999999	999999	999999	999999	999999
H	999999	999999	999999	999999	999999	999999	112	0	999999	999999	999999	999999	135	999999	999999	999999	999999	999999	999999	999999	999999	999999	999999	999999
I	999999	999999	999999	999999	999999	999999	999999	70	0	999999	999999	999999	999999	999999	999999	999999	999999	999999	999999	999999	999999	999999	999999	999999
J	999999	999999	999999	999999	999999	999999	999999	999999	75	0	999999	999999	102	999999	999999	999999	999999	999999	999999	999999	999999	999999	999999	999999
K	433	999999	999999	999999	999999	999999	999999	999999	999999	999999	0	183	999999	999999	999999	999999	999999	999999	153	999999	999999	999999	999999	999999
L	999999	999999	999999	999999	999999	999999	999999	999999	999999	999999	183	0	92	112	999999	999999	999999	999999	124	999999	999999	999999	999999	999999
M	999999	999999	999999	999999	999999	999999	115	999999	999999	999999	999999	85	0	120	999999	999999	999999	999999	999999	999999	999999	999999	999999	999999
N	999999	999999	999999	999999	999999	999999	87	999999	999999	999999	999999	999999	112	0	999999	999999	999999	999999	999999	999999	999999	999999	999999	999999
O	999999	999999	999999	999999	388	999999	999999	999999	999999	999999	999999	999999	999999	999999	0	999999	999999	999999	999999	999999	999999	999999	999999	999999
P	999999	999999	999999	999999	85	999999	999999	999999	999999	999999	999999	999999	999999	999999	999999	0	999999	999999	999999	999999	185	999999	999999	999999
Q	999999	999999	999999	999999	999999	999999	999999	999999	999999	999999	999999	999999	62	999999	0	999999	999999	999999	999999	999999	155	999999		
R	999999	999999	999999	999999	999999	999999	999999	999999	999999	999999	999999	999999	999999	85	0	999999	999999	999999	999999	999999	999999	999999	999999	999999
S	999999	999999	999999	999999	999999	999999	999999	999999	999999	75	999999	999999	999999	999999	999999	999999	0	999999	999999	999999	999999	999999	999999	999999
T	999999	999999	999999	999999	999999	999999	999999	999999	999999	999999	999999	999999	999999	999999	999999	999999	999999	0	105	999999	160	999999		
U	999999	999999	234	999999	999999	999999	999999	999999	999999	999999	999999	999999	999999	999999	999999	999999	195	0	999999	999999	38			
X	999999	999999	999999	999999	999999	999999	999999	999999	999999	999999	999999	999999	185	999999	999999	999999	999999	999999	0	85	999999			
Y	999999	999999	999999	999999	999999	999999	999999	999999	999999	999999	999999	999999	999999	999999	999999	160	999999	85	0	144				
Z	999999	999999	215	999999	999999	999999	999999	999999	999999	999999	999999	999999	999999	93	999999	999999	38	999999	144	0				

The aim of our study is to determine the number of vehicle per segment, we start by calculating the volume of vehicles per route, this is done by multiplying each column (node) of the vehicle flow vector, by each row of the matrix that defines the transit probabilities (we assume that each driver knows previously the shortest path to use in the study area and just a few of them ignore the shortest itinerary) between nodes. The probability matrix is defined as being in a node for example A, which will be the ability to address the remains of the nodes, such that the high probabilities are for the neighboring nodes and then according to the location within the graph and its proximity to principal's streets, we resume the result in Table 2.

Table 2. Volume of vehicles per route and transit probabilities

The transit probabilities between nodes

	A	B	C	D	E	F	G	H	I	J	K	L	M	N	O	P	Q	R	S	T	U	X	Y	Z
A	0	0.17	0.03	0.04	0.12	0.04	0.02	0.02	0.03	0.03	0.21	0.05	0.02	0.04	0.04	0.03	0.01	0.01	0.04	0.01	0.01	0.01	0.01	0.01
B	0.13	0	0.18	0.08	0.14	0.02	0.02	0.02	0.04	0.03	0.01	0.02	0.07	0.03	0.02	0.03	0.02	0.02	0.04	0.02	0.01	0.01	0.01	0.02
C	0.09	0.17	0	0.12	0.11	0.04	0.02	0.01	0.02	0.02	0.04	0.03	0.02	0.03	0.09	0.03	0.01	0.02	0.02	0.03	0.02	0.03	0.01	0.02
D	0.02	0.06	0.16	0	0.24	0.03	0.01	0.01	0.01	0.01	0.01	0.02	0.01	0.02	0.06	0.04	0.07	0.14	0.01	0.02	0.01	0.01	0.01	0.02
E	0.01	0.02	0.05	0.12	0	0.07	0.01	0.01	0.01	0.01	0.06	0.11	0.02	0.08	0.14	0.1	0.02	0.01	0.01	0.01	0.09	0.01	0.01	0.01
F	0.02	0.01	0.02	0.11	0.22	0	0.02	0.01	0.01	0.02	0.05	0.07	0.05	0.21	0.08	0.02	0.01	0.02	0.01	0.01	0.01	0.01	0.01	0.01
G	0.01	0.01	0.02	0.08	0.17	0.24	0	0.02	0.02	0.02	0.05	0.04	0.06	0.09	0.05	0.01	0.02	0.02	0.01	0.01	0.01	0.01	0.01	0.01
H	0.02	0.03	0.02	0.04	0.1	0.11	0.12	0	0.01	0.01	0.02	0.02	0.02	0.05	0.25	0.05	0.01	0.03	0.01	0.01	0.01	0.04	0.01	0.01
I	0.01	0.02	0.02	0.04	0.06	0.04	0.05	0.22	0	0.02	0.05	0.03	0.07	0.02	0.01	0.02	0.01	0.01	0.01	0.01	0.01	0.04	0.01	0.01
J	0.01	0.01	0.02	0.04	0.07	0.04	0.04	0.07	0.23	0	0.06	0.02	0.12	0.07	0.08	0.02	0.01	0.02	0.02	0.01	0.01	0.01	0.01	0.01
K	0.16	0.02	0.02	0.02	0.05	0.02	0.03	0.04	0.06	0.07	0	0.21	0.03	0.03	0.04	0.01	0.01	0.01	0.01	0.01	0.01	0.01	0.01	0.01
L	0.01	0.02	0.03	0.05	0.07	0.06	0.02	0.01	0.02	0.03	0.17	0	0.08	0.15	0.05	0.02	0.02	0.02	0.12	0.01	0.01	0.01	0.01	0.01
M	0.01	0.01	0.02	0.05	0.08	0.11	0.19	0.06	0.04	0.03	0.02	0.03	0	0.16	0.07	0.02	0.01	0.02	0.02	0.01	0.01	0.01	0.01	0.01
N	0.02	0.01	0.02	0.02	0.07	0.19	0.03	0.01	0.02	0.02	0.09	0.16	0.13	0	0.06	0.02	0.01	0.02	0.03	0.02	0.02	0.01	0.01	0.01
O	0.02	0.03	0.1	0.12	0.2	0.02	0.01	0.01	0.01	0.01	0.05	0.05	0.02	0.06	0	0.08	0.01	0.03	0.01	0.01	0.03	0.03	0.01	0.01
P	0.01	0.03	0.07	0.07	0.16	0.08	0.01	0.01	0.01	0.02	0.02	0.02	0.03	0.07	0	0.05	0.05	0.01	0.02	0.02	0.18	0.02	0.02	
Q	0.01	0.02	0.02	0.03	0.09	0.07	0.01	0.01	0.01	0.01	0.02	0.01	0.01	0.01	0.05	0.22	0	0.01	0.01	0.06	0.05	0.09	0.17	0.01
R	0.01	0.02	0.02	0.03	0.07	0.03	0.01	0.01	0.01	0.01	0.01	0.02	0.01	0.06	0.03	0.04	0.13	0.21	0	0.01	0.06	0.04	0.09	0.07
S	0.01	0.02	0.02	0.04	0.07	0.02	0.02	0.08	0.11	0.26	0.01	0.03	0.06	0.03	0.08	0.03	0.01	0.02	0	0.02	0.02	0.01	0.01	0.02
T	0.02	0.1	0.15	0.02	0.05	0.02	0.01	0.01	0.01	0.01	0.02	0.01	0.01	0.06	0.03	0.09	0.06	0.01	0.03	0	0.22	0.03	0.04	0.07
U	0.01	0.03	0.17	0.02	0.06	0.02	0.01	0.01	0.01	0.01	0.02	0.02	0.01	0.02	0.04	0.03	0.03	0.11	0.01	0.1	0	0.05	0.07	0.14
X	0.01	0.02	0.04	0.02	0.06	0.02	0.01	0.01	0.01	0.01	0.01	0.02	0.01	0.02	0.05	0.21	0.04	0.03	0.01	0.08	0.04	0	0.16	0.11
Y	0.01	0.02	0.03	0.03	0.09	0.02	0.01	0.01	0.01	0.01	0.01	0.01	0.01	0.07	0.02	0.03	0.03	0.01	0.13	0.09	0.11	0	0.23	
Z	0.01	0.05	0.09	0.02	0.07	0.02	0.01	0.01	0.01	0.01	0.02	0.02	0.01	0.02	0.05	0.03	0.04	0.12	0.01	0.03	0.19	0.02	0.14	0

Volume of vehicles per route

	A	B	C	D	E	F	G	H	I	J	K	L	M	N	O	P	Q	R	S	T	U	X	Y	Z
A	0	150	62	8	8	26	35	35	168	17	17	8	35	8	17	26	35	35	44	35	35	26		
B	165	0	176	66	66	11	11	22	22	99	22	11	11	55	11	22	33	44	55	99	22	22	33	
C	94	215	0	107	108	13	26	13	13	53	40	26	40	107	26	26	40	13	26	174	40	40	80	
D	31	95	158	0	127	42	10	10	10	10	31	31	10	21	42	84	10	21	42	31	42	42		
E	30	107	122	153	0	138	15	15	15	15	92	92	15	122	153	138	30	46	15	15	46	122	15	15
F	18	46	46	101	148	0	9	9	18	18	55	64	27	138	92	46	9	18	9	9	9	9		
G	3	12	14	14	18	25	0	25	1	1	5	5	5	5	12	10	1	3	1	1	3	1	1	1
H	5	20	25	25	50	55	60	0	5	5	10	10	10	20	110	25	5	15	5	5	5	20	5	5
I	5	11	11	23	29	23	52	129	0	11	23	17	29	11	117	23	5	11	5	5	5	17	5	5
J	5	10	10	20	25	25	35	65	65	0	30	10	55	20	65	10	5	10	10	5	5	5	5	5
K	193	48	32	32	80	80	48	96	113	113	0	176	96	96	80	16	16	16	192	16	16	16	16	16
L	39	59	39	39	79	68	19	19	19	29	119	0	89	109	49	29	19	19	79	9	9	9	9	9
M	3	6	9	12	15	34	49	15	37	3	6	6	0	59	21	6	3	3	3	3	3	3	3	3
N	42	42	33	25	75	101	25	8	16	16	84	117	42	0	50	25	16	25	25	16	16	8	8	
O	20	71	81	81	163	51	20	10	10	10	61	61	30	61	0	71	10	30	10	10	40	71	20	20
P	8	43	69	69	104	60	8	8	8	17	17	17	25	60	0	25	34	8	17	25	120	69	34	
Q	2	5	5	7	18	15	2	2	2	2	2	13	52	0	2	2	15	13	23	44	15			
R	3	6	6	6	20	10	3	3	3	3	6	6	10	10	13	43	70	0	3	20	13	26	23	20
S	11	16	16	22	11	11	11	49	66	99	5	16	44	27	49	11	5	11	0	5	16	16	11	16
T	9	45	68	9	13	9	4	4	4	9	4	4	4	27	13	13	27	4	0	100	18	18	36	
U	5	41	83	11	17	11	5	5	5	11	11	5	11	23	17	17	64	5	58	0	29	41	94	
X	10	26	26	10	21	10	5	5	5	15	10	5	10	31	63	21	31	5	21	31	0	89	63	
Y	3	14	14	11	18	7	3	3	3	3	3	3	3	26	7	11	11	3	44	33	52	0	81	
Z	3	31	31	7	15	7	3	3	3	7	7	3	7	15	7	11	47	3	11	59	39	55	0	

Our objective is to know the volume of vehicles per segment (segment is arc of a graph), we use Dijkstra algorithm to calculate the shortest path from each node to the other nodes of the graph (algorithm 1), and then we use this result with the matrix defined in Table 2 (volume of vehicles per route or itinerary) to add the vehicles that appear in the same arc (algorithm 2).

The result is the number of vehicles circulating in each segment of the graph.

To carry out this task we apply the first Wardrop principle, the user-optimal equilibrium [15], along with Dijkstras Algorithm to define the minimum itinerary from each node [16]:

Algorithm 1 Dijkstra's Algorithm to define the minimum itinerary from each node

---
dist[s] ←0
for all v ∈ V–{s}
  do dist[v]←∞
    S←∅
    Q←V
    while Q≠∅
      do u ←mindistance (Q, dist)
        S←S∪{u}
          for all v ∈ neighbors[u]
        do if dist[v]> dist[u]+ w (u, v)
        then d[v] ←d[u]+ w (u, v)
          return dist

The Dijkstra algorithm returns the shortest path from one node to each node in the graph, as depicted in Fig. 4.

The result can be represented as:

```
0 --> 0 = 0 meters.
0 --> 1 --> = 720 meters.
0 --> 1 --> 2 --> = 892 meters.
0 --> 10 --> 11 --> 13 --> 5 --> 4 --> 3 --> = 1063 meters.
......
23 --> 20 --> 19 --> = 233 meters.
23 --> 20 --> = 38 meters.
23 --> 22 --> 21 --> = 229 meters.
23 --> 22 --> = 144 meters.
23 --> 23 = 0 meters.
```

Fig. 4. Dijkstra algorithm result

Now, to calculate the volume of vehicles per segment, we proceed by using the algorithm 2 witch combine the result of the Dijkstra's algorithm depicted in Fig. 4, along with the number of vehicles per route as illustrated in Table 2.

Algorithm 2 is based on the singly linked list to stack the paths given by Dijkstra's algorithm. The linked list is a dynamic data structure used to implement stacks, queues and graphs. The structure is consisting of a series of nodes of the same type which are linked together by pointers. Comparing to static data structures, such as arrays, dynamic data structures have the flexibility to grow and shrink in size during program execution. The linked list has the advantage is reducing the access time, inserting and deleting data to the list without reallocation of the entire list [17].

Algorithm 2 Calculate the number of cars that will go through a segment 'S'

Input: N paths returned by dijkstra algorithm
Volume of vehicles per route (Volume_Route)
Segment S : Origin_Node, Destination_Node
Output: SumCars=0, number of cars that will go through the S segment
for k=0 to n **do**
    Fill list[k] with the linked list that contains index of each node in path k
end for
for k=0 to n do
if Search_Segment_Linked_List (list[k], Origin_Node, Destination_Node) is found in path k
    then
        SumCars=Volume_Route(list[k].get_Head()),(list[k]. get_Tail ()) + Sum-Cars;
    end if
end for
return SumCars

2.2 Results and Discussion

We have resorted to the formula (1) described above what is a multiplication result of, volume of vehicles per segment, the distance traveled and the emission factor.

    We have previously identified the volume of vehicles per route, and we have used the Dijkstra's algorithm (algorithm 1) to calculate the shortest paths between nodes in the graph. Using these results together with the volume of vehicles per route (algorithm 2), we can calculate the volume of vehicles for each segment, as shown in Table 3.

Table 3. Volume of vehicles per segment

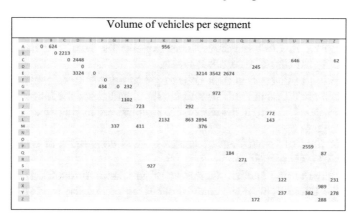

	A	B	C	D	E	F	G	H	I	J	K	L	M	N	O	P	Q	R	S	T	U	X	Y	Z
A	0	624											956											
B		0	2213																					
C			0	2448																	646			62
D				0														245						
E				3324	0									3214	3542	2674								
F						0																		
G						434	0	232																
H														972										
I								1102																
J									723			292												
K																		772						
L											2132	863	2894					143						
M							337	411					376											
N																								
O																								
P																					2559			
Q														184										87
R															271									
S									927															
T																								
U																			122					231
X																						989		
Y																			237	382				278
Z														172						288				

With this matrix that describes the volume of vehicles in each segment of the graph, we can calculate the quantity of CO2 emitted in each segment, with the amount emitted by vehicle in the rush hour of the day is 352 gr/Km [18]. Consequently, we compare the different streets to see which contributes the most in the pollution within this zone. As illustrated in the Fig. 5, the emissions expressed in gr/km in rush hour in each street in the area, indicating with a red circle the high emission rates.

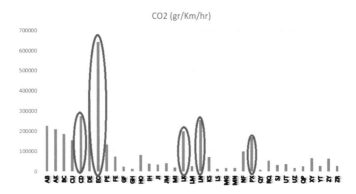

Fig. 5. CO2 emissions in different streets

Our first observation is that exist a big difference between segment emissions, the sections that contribute the most in CO2 emissions are the streets around the node E: EO, CD, LN, EP but the most significant is EO.

Access to the city center can be done by three roads, EO which is the short route (causing congestion) and the long route EP or EK.

DE, EO, is the avenue that contains the most pedestrians, because there are two schools at the side of a roundabout E and it's also an access to the city center, this is for the vehicles using the direction of circulation D -> E -> O. The other direction of circulation O -> E -> D, is taken by the majority of vehicles to access to the university, hospital, and other destinations.

According to the first Wardrop principle, almost drivers choose the shortest route to minimize the cost of each user to access to the city center (EO in the graph), which causes an increase in CO2 emissions. This explains the results of the Fig. 4, CO2 emissions are higher on EO than other streets.

Another result to discuss, in rush hour to get to node B or C, most drivers avoid passing through node E, that contains a traffic light, and chose the longer path K-A-B-C, and this explains the high degree of CO2 emissions in this route AB and AK, according to Fig. 4.

The traffic light of roundabout E, causes a great congestion at rush hour in the avenues, EO, ED, EP and EL.

Then we have to find another alternative of circulation through this area of study and mainly the traffic light of the roundabout E, to reduce the emission rate due to urban transport.

As we can see, it is a very large amount of emissions in this region and that will surely cause a health degradation of citizens. To reduce CO2 emissions, it is interesting to introduce new urban mobility planning mechanisms, using sustainable urban mobility plans [19]. But to carry out this, it is necessary to find alternatives of circulation, also the parking of the cars decreases the capacity of circulation in streets and therefore it produces a great congestion.

3   Conclusion

Since the signing of the Paris Climate Agreement (COP 21), commitments to reduce greenhouse gas emissions are increasing.

In this paper, we have developed a methodology to quantify carbon dioxide emissions in an area of the city of Tangier, detailing those emissions in each street. We have quantified the relation between the number of vehicles circulating and the CO2 emissions, in this area the health of citizen will be affected by the high level of CO2 emissions, especially children, and drastic decision should be taken to avoid dramatic degradation in health.

Emissions due to transport are increasingly influential in large cities, citizens are also sensitive to environmental problems and global warming whose consequences affect them directly. To reduce this effect, we must reduce the dependence with the private car and promote the transport in common also ensuring intermodality.

As a perspective, we can study how to find circulation alternatives simulating different solutions and using the second Wardrop principles [15].

References

1. Rupprecht Consult. The state of the art of sustainable urban mobility plans in Europe, Edinburgh Napier University, September 2012
2. Blanc, L.: La lutte contre le rechauffement climatitique: une croisade absurde, couteuse et inutile, societe de calcul mathematique SA, Aout 2015
3. Alexopoulos, A., Assimacopoulos, D., Mitsoulis, E.: Model for traffic emissions estimation. Atmos. Environ. **27B**(4), 435–446 (1993)
4. Eerens, H.C., Sliggers, C.J., Van Den Hout, K.D.: The CAR model: the Dutch method to determine city street air quality. Atmos. Environ. **27B**(4), 389–399 (1993)
5. Xia, L., Shao, Y.: Modelling of traffic flow and air pollution emission with application to Hong Kong island. Environ. Model. Softw. **20**, 1175–1188 (2005)
6. Gois, V., Maciel, H., Noriega, L., Almeida, C., Torres, P., Mesquita, S., Ferreira, F.: A detailed urban road traffic emissions inventory model using aerial photography and GPS surveys. In: EPA's 16th Annual International Emission Inventory Conference, 16 May 2007
7. Ahn, K., Rakha, H.: The effects of route choice decision on vehicle energy consumption and emissions. Transp. Res. Part D Transp. Environ. **13**, 151–167 (2008)
8. Gostadi, M., Rossi, R., Gecchele, G., Lucia, L.D.: Annual average daily traffic estimation from seasonal traffic counts. Procedia – Soc. Behav. Sci. **87**, 279–291 (2013)

9. Fu, M., Kelly, J.A., Clinch, J.P.: Estimating annual average daily traffic and transport emissions for a national road network: a bottom-up methodology for both nationally-aggregated and spatially-disaggregated results. J. Transp. Geogr. **58**, 186–195 (2017)
10. Bharadwaj, S., Ballore, S., Rohit, R., Chandel, K.: Impact of congestion on greenhouse gas emissions for road transport in Mumbai metropolitan region. In: World Conference on Transport Research, WCTR 2016, Shanghai, 10–15 July 2016
11. Bisbe, S., Raffenel, D., Beauzamy, B.: Vitesse des vehicules et emissions de co2. Societe de calcul mathematique SA, outils d aide a la decision, aout 2013
12. Grote, M., Willaims, I., Preston, J., Kemp, S.: Including congestion effects in urban road traffic co2 emissions modeling: do local government authorities have the right option? Transp. Res. Part D **43**, 95–106 (2016)
13. http://www.4c.ma/fr
14. Sheffi, Y.: Urban Transportation Networks. Massachusetts Institute of Technology, Prentice-Hall, Englewood Cliffs (1985)
15. Wardrop, J.G.: Some theoretical aspects of road traffic research. In: Proceeding of the Institute of Civil Engineers (1953)
16. http://math.mit.edu/~rothvoss/18.304.3PM/Presentations/1-Melissa.pdf
17. Shyram: Multiply linked lists. http://www.classle.net/projects/multiply-linked-lists. Accessed 7 May 2012
18. http://www.bilans-ges.ademe.fr/static/documents/[Base%20Carbone]%20Documentation%20générale%20v11.0.pdf
19. Sustainable urban transport planing SUTP Manual, Guidance for stakeholders (2007)

Exploring Apache Spark Data APIs for Water Big Data Management

Nassif El Hassane[(⊠)] and Hicham Hajji

School of Geomatic Sciences and Surveying Engineering,
SGIT, IAV Institute, Rabat, Morocco
nassif.hassane@gmail.com, h.hajji@iav.ac.ma

Abstract. Managing data complexity is a recurrent problem in multiple domains related to water resources management such as utilities, hydrological and meteorological modelling. Recently and since the advent of intelligent sensors, we observe a systemic growth in the volume of collected data. Besides, these kinds of sensors generate near real-time data under various formats. To get the right value of this kind of water datasets we need to design new solutions, efficient enough to manage massive data coming from intelligent sensors in near real time and under various formats. We present in our paper a reference architecture for managing massive data collected from smart meters. Also, we show how recent advances in big data technologies mainly the Apache Spark project can effectively be used to obtain insights from massive datasets. Finally, we will focus on presenting the advantages that provide the distributed execution model of Spark by exploring three Apache Spark APIs: RDD, Dataframe, and SparkR.

Keywords: Big Data · Spark · Water management · RDD · Dataframe

1 Introduction

During the past few years, growing data volumes has been considered as a key challenge for organizations pushing them to look for new approaches to scale their applications and computations. One of the solutions they considered was to distribute data storage and processing across clusters of hundreds of machines (Example of Google, Facebook, Amazon …). In addition to simple queries, complex algorithms like machine learning and graph analysis are becoming common in many domains. Also streaming analysis of real-time data is required to let organizations take timely action. Water management is not an exception since data collection and processing is becoming a challenge for practitioners, IT teams, and decision makers. Either it was for managing river basin information, for managing Water utilities data, or for carrying out data intensive hydrologic modelling, the data management task has always been challenging. And has become more difficult with the advent of real time sensors, remote imagery, and the need to speed up the process of decision making [5–7]. Throughout this paper, we will present a reference architecture for handling and managing smart metering water datasets. We demonstrate how recent advancement in Big Data Technologies (especially Apache Spark project) can handle water big data

© Springer Nature Switzerland AG 2019
M. Ezziyyani (Ed.): AI2SD 2018, AISC 913, pp. 105–117, 2019.
https://doi.org/10.1007/978-3-030-11881-5_10

efficiently with fault tolerance for getting insights from those datasets. Finally, we will highlight the advantages that provide the distributed execution model of Spark by exploring three APIs and abstractions provided by Apache Spark: RDD, Dataframe, and SparkR. The aim of this paper is mainly to explore how Spark can be used with different abstractions to handle Big data constraints as encountered in Smart Metering Data processing. Due to lack of space, the impact of volume on such approaches will not be addressed in this paper, and will be developed in a further work.

2　Big Data Examples in Water Management

2.1　Smart Grid and Water Smart Metering

With the challenges of managing water resource scarcity, and for preventing man-made disasters like natural flooding, and minimizing the impacts of drought in arid regions, Water utilities need to reinvent their monitoring techniques and to rely heavily on IT technologies. Such Newly techniques can play central role for assisting Water decision makers by giving them faster access to better information to make better decisions and to rapidly disseminate that information to customers and other stakeholders. Among the recent advances, one can list the two following initiatives and technologies:

- Advanced metering systems: Which are systems that enable measurement of detailed, time-based information and frequent collection and transmittal of such information to various parties [3]. All the data is collected from one smart meter as the water enters the property. This can measures water usage, pressure and temperature with extreme accuracy measurements taken each period of time (second, minute … etc.) [8]. This collected water consumption flow data can enhance many tasks such as leaks detection and understanding end use events (e.g. shower, toilet, washing machine, etc.) [9].
- Real Time Sensors for specific measurements have been developed recently for Water quality monitoring [1], and they aim to simplify remote water quality monitoring. With multiple sensors that measure a dozen of the most relevant water quality parameters, they are suitable for potable water monitoring, chemical leakage detection in rivers, remote measurement of swimming pools and spas, and levels of seawater pollution.

2.2　Data Intensive Hydrologic Modelling

The computer models of watershed hydrology are highly data intensive and the time consumed running hydrologic models (especially physically based and distributed hydrologic models) is still a concern for hydrologic practitioners and scientists [2, 4]. In addition, the complexity of the calibration problem has increased substantially. Recall that the successful application of a hydrologic model depends on how well the model is calibrated.

2.3   Remote Sensing, Atmospheric Measurements and Climate Change Modelling

Many climate models are using high-end supercomputers to complete a comprehensive set of climate change simulations that will be used to advance scientists' knowledge of climate variability and climate change. In addition, remote sensing observations (e.g., remote sensing imagery, Atmospheric Radiation Measurement (ARM) data) are generating large amounts of scientific data (see Table 1 that shows some of the climate and earth systems data stored at the Earth System Grid (ESG) portal).

Table 1. Example of scientific data stored on the earth system grid

	CIMP5	ARM	DACC
Sponsor	SciDAC	DOE/BER	NASA
Description of data	40+ Models	Atmospheric processes and cloud dynamics	Biogeochemical dynamics, FLUXNET
Archive size	~6 PB	~200 TB	1 TB
Year started	2010	1991	1993

3   Overview of Constraints in Big Data Water Management Datasets

Water datasets, whether captured through remote sensors or large-scale simulations, as stated above, has always been Big. Similarly to traditional Big Data, three features can mainly characterize it: Volume, Variety, and Velocity, defined as three "V" dimensions by Gartner in 2001 [13]:

3.1   Velocity

The velocity of big data in Water Data management involves the generation of data at a rapid growth rate, and efficiency of data processing and analysis. The data should be analyzed in a near-real-time manner to achieve a given task, e.g.: flooding prediction and leak detection.

3.2   Variety

In terms of variety, Water-Dataset data consist of multisource (sensors, smart metering, hydrological, DEM, etc.), at different resolutions (temporal, and spatial).

3.3   Volume

To illustrate the growth of data volume generated by Water Management, the following Table 2 shows how the data collected from smart metering evolves according to the granularity of measurement.

Table 2. Data storage issues for smart metering [14]

Granularity	Number of measures by year	Storage per year	Storage per 40 million per year
One second	31 536 000	120 MB	4 PB
One minute	525 600	2 MB	76 TB
One hour	8760	34 KB	1 TB
One day	365	1 KB	54 GB

4   Big Data Architecture for Managing Massive Water Datasets Using Spark

As stated earlier, we will focus in this paper on managing water big data coming from Smart Metering datasets with efficiency, scalability and fault tolerance.

In order to achieve those purposes, it is not enough to focus only on the processing layer only. We should rather treat all layers, from data collection, ingestion, processing and even visualization. It is in this sense that we propose an end-to-end architecture (Fig. 1) based on big data tools to ensure timely collection, rapid ingestion and efficient query processing.

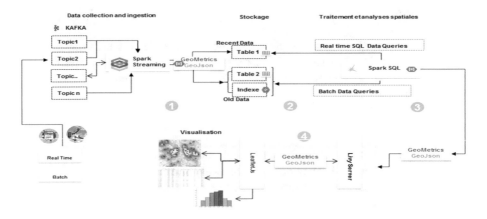

Fig. 1. End to end big data architecture for managing massive water datasets.

This architecture (Fig. 1) is able respond to both usual use cases: near real time and batch. Data collection and ingestion are carried out by making use of three ingestion tools: using Apache Camel [15] for files data sources and then through Apache Scoop [16] for DBMS data sources and finally through Apache Kafka [17] for streaming data (real water smart metering datasets). Kafka is a subscriber/consumer messaging system, horizontally scalable and fault-tolerant. Data stored in Kafka is then consumed by Apache Spark that can execute cleaning, transformation and processing of data before sending them to apache Cassandra [17]. Once the data is stored in Cassandra, Water smart metering data are made available to user through Spark SQL, one of Spark's APIs. One of analytical queries that can be seamlessly executed within our architecture are: leaks detection and customer profiling queries[1].

As smart metering data can also be defined as data source with spatial component such as customer coordinates (X, Y) or smart metering locations (latitude, longitude). We have introduced in our architecture at least one case where locations are handled: During ingesting phase of smart metering data into Cassandra, spatial indexes (such as Z-Index) can be constructed based upon coordinates and location.

4.1   Processing Water Datasets Using Apache Spark

In this part, three different approaches based on Apache Spark for processing Water Smart Metering has been explored. Recall that Apache Spark is an open source parallel data flow system built on the concept of Resilient Distributed Datasets (RDD), which is a fault-tolerant collection of elements that can be operated on in parallel [10–12]. Because RDDs are cached in memory and the data flow is created lazily, Spark's model is well suited for bulk iterative algorithms.

In the following sections, we will briefly recall the three Apache Spark abstractions, RDD, Dataframe and SparkR, that will be explored in our work.

Resilient Distributed Dataset (RDD)

The first abstraction of Apache Spark we will explore is the Resilient Distributed Dataset (RDD) that represents an immutable, partitioned collection of elements that can be operated on in parallel.

Internally, each RDD is characterized by five main properties (see Fig. 2). RDD supports two types of operations (Table 3):

Fig. 2.   Resilient distributed dataset properties.

- Transformations, which create a new dataset from an existing one.
- Actions, which return value to the driver program after running a computation on the dataset.

Table 3. Two types of operations supported by Spark.

Operation	Function
RDD transformation	Map(func); flatmap() Filter(func); mappartitions(func) Mappartitionwithindex() Union(dataset); intersection(other-dataset) Distinct(); groupbykey() Reducebykey(func, [numtasks]) Sortbykey(); join();coalesce()
RDD action	Count(); collect() Take(n); top() Countbyvalue() Reduce(); fold() Aggregate(); foreach()

All transformations in Spark are lazy, they do not compute their results right away. Instead, they just remember the transformations applied to some base dataset (e.g. a file). The transformations are only computed when an action requires a result to be returned to the driver program.

Spark SQL Dataframe

Spark SQL is an extension of Apache Spark for structured data processing. Unlike the basic Spark RDD API, it provides Spark with more information about the structure of both the data and the computation being performed and it allows executing SQL queries written using a basic SQL syntax.

Table 4. RDD and dataframe comparison.

	RDD	DataFrame
Data formats	Can be used to process structured as well as unstructured data	Data is organized into named columns
Data representations	Is a distributed collection of data elements spread across many machines over the cluster. They are a set of Scala or Java objects representing data	Data frame data is organized into named columns. Basically, it is as same as a table in a relational database
Optimization	There was no provision for optimization engine in RDD	By using Catalyst Optimizer, optimization can takes place in dataframes
Serialization	Spark uses java serialization, whenever it needs to distribute data over a cluster	In dataframe, we can serialize data into off-heap storage in binary format
Efficiency and memory use	When serialization executes individually on a java and scala object, efficiency decreases	Use of off-heap memory for serialization reduces the overhead

DataFrame can be considered as a distributed collection of data organized into named columns and can be constructed from a wide array of sources such as: structured data files, tables in Hive, external databases, or existing RDDs. To understand the main differences between Apache Spark RDD and DataFrame we compare them on the basis of different features (Table 4).

SparkR Package Using R Software
R is a popular tool for statistics and data analysis. It gives rich visualization capabilities and a large collection of libraries. However, in the context of Big Data, it is designed only to run on in-memory data, which makes it unsuitable for large datasets.

SparkR is an R package that provides a frontend to use Apache Spark cluster from R (Fig. 3). It provides a distributed data frame implementation (Similar to R dataframes) that supports operations like selection, filtering and aggregation but on large datasets.

SparkR allows users to connect R programs to a Spark cluster from RStudio, R shell, Rscript or other R IDEs. It can operate on a variety of data sources through the Spark DataFrame interface.

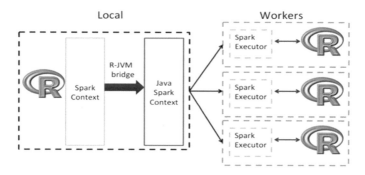

Fig. 3. Diagram of SparkR connecting to spark cluster (https://aws.amazon.com/fr/blogs/big-data/crunching-statistics-at-scale-with-sparkr-on-amazon-emr/).

4.2 Three Approaches for Processing Water Datasets: RDD, Dataframe and SparkR

To illustrate the use of Apache Spark for Water Management, we will present briefly how Spark ecosystem can be used to process from simple to complex analytical queries on water smart metering datasets. Practically, we explored Spark processing by making use of three approaches:

- Using Resilient Distributed Datasets as a central backbone for describing our queries.
- Using DataFrame and Spark Sql for representing and formulating our queries.
- Using SparkR for interacting with our Water Smart Metering Datasets using R API over Apache Spark.

The code related to the three approaches can be found in https://github.com/hajjihi/BD4WM. Data used in our prototype are composed of 2246 files downloaded from Smart Metering Information Portal SMIP[2]. SMIP is an online environment that allows researchers to collect, preserve, access, and collates data gathered from smart meter devices. It is also a secure service provided to assist researchers from the Smart Water Research Centre in the query and maintenance of water logger details and associated data (logger data, household survey data, logger history).

Using Spark Dataframe

Water Smart Metering Data Ingestion
When Spark starts ingesting logger data files, available through any convenient storage such as HDFS or a local file system (available on all nodes), it constructs an RDD collection. Starting from this point, data are presented as a distributed dataset that can be operated on in parallel.

Preparing Case Classes for Inferring Schema
Case classes are regular classes that provide a recursive decomposition mechanism via pattern matching. In our case, the case class describes granular information collected from smart metering such as:

- Id of the Smart meter.
- Date of the measure (year, month, day, and interval).
- Value of the measure.

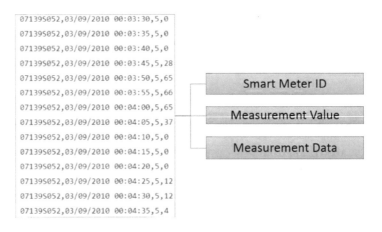

```
case class SmartWaterMeasure(Meter: String,
                             year: Int,
                             month:Int, day:Int,
                             Interval:Int,
                             Value:Int)
```

---
[2] https://code.google.com/p/smart-meter-information-portal/.

Then comes the data preparation phase where Spark maps Smart metering data into the above case class SmartWaterMeasure, and transforms it to Dataframe.

```
val SmartMeter = c.textFile("data").map(_.split(","))
.map(p ⇒ SmartWaterMeasure(p(0), p(1).substring(6,10)
.toInt , p(1).substring(3,5).toInt, p(1).substring(0,2)
.toInt, p(2).trim.toInt ,p(3).trim.toInt)).toDF()
SmartMeter.registerTempTable("SmartMeter")
```

The Scala interface for Spark SQL supports automatically converting an RDD containing case classes to a DataFrame. Once The RDD is implicitly converted to a DataFrame and then be registered as a table, it can be used in subsequent SQL statements.

Analytical Queries for Smart Metering Datasets
To illustrate the use of our approaches, we will present some analytical queries and show how they can be expressed using the Dataframe abstraction.

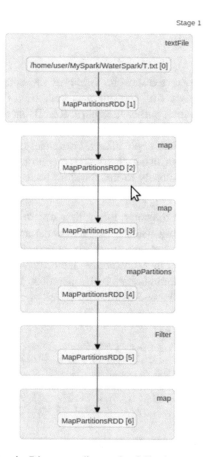

Fig. 4. Direct acyclic graph of filtering query.

– Query 1: Getting water smart meter data with measurement year greater than 2009

The first query is a simple one that returns Water Smart Meter data with measure attribute year greater than 2009. In the corresponding Directed Acyclic Graph of the Spark query (Fig. 4), we can read that the job associated with the query is associated with a chain of RDD dependencies organized in a direct acyclic graph (DAG).

First, it performs an ingestion operation of available logger files, and then use the Map operation before applying the Filter operation. Recall that for tuning and optimization, Spark uses the Project Tungsten to improve the efficiency of memory and CPU.

As the query is just about just filtering smart metering dataset, there is no shuffle needed between nodes of the cluster. The corresponding DAG shows consequently a single stage composed of subsequent tasks.

– Query 2: Aggregating smart meter data

This example of aggregation queries tries to group smart metering data with two attributes: Meter Id, and Year of Measurement, and then apply two aggregates methods: Average and Maximum.

```
SmartMeter.groupBy(("Meter"),("year")).agg(avg(("Value")),
                                            max(("Value")))
```

In the corresponding Directed Acyclic Graph of the Spark query (DAG), we can notice that the job associated with the query is composed of two stages because of shuffling due to the aggregation part of the query. Recall that Shuffling is the process of data transfer between stages. It is one of the problems that need to be minimized and tuned when developing Big Data applications. Fortunately, the most part of shuffling is taken into consideration when using Spark SQL, contrary to RDD based approach, where user should take care closely to the shuffling issue (Fig. 5).

Using Spark RDD
For this case, we found that the most interesting way to use RDD for constructing analytical queries is to making use of the Accumulator variable (see code below). Accumulators are variables that are used for aggregating information across the executors. Similar to counters in MapReduce, they are variables that are "added" to through an associative and commutative "add" operation. They are designed to be used safely and efficiently in parallel and distributed Spark computations.

```
val resultMonth = Bymois.combineByKey(
    (v) ⇒ (v, 1),
    (acc: (Int, Int), v) ⇒ (acc._1 + v, acc._2 + 1),
    (acc1: (Int, Int), acc2: (Int, Int)) ⇒
    (acc1._1 + acc2._1, acc1._2 + acc2._2)
    ).map{ case (key, value) ⇒
    (key, value._1 / value._2.toFloat) }
    resultmois.collectAsMap().map(println(_))
```

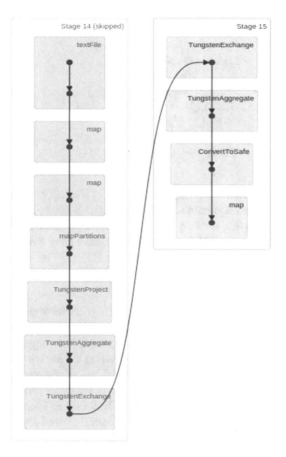

Fig. 5. Direct acyclic graph of aggregation query.

The above code computes the average of water consumption by month and by customer. First, each value are transformed to pairRDD (value, 1). Then to compute the Average-by-key, the map method is used to divide the sum by the count for each key. Finally, CollectAsMap method is used to return the average-by-key as a dictionary. Keys in this example are month and customer.

Let's recall that even Accumulator, as shared variable, can help expressing analytical queries, but it suffers from some issues such as the important number of the accumulators copies that each executor should handle. Also handling manually shuffling for advanced queries makes RDDs very difficult to use and to maintain.

Using SparkR

Creating connection between R software and Spark cluster can be summarized by ingesting data from textFile Spark action and constructing a dataframe (in the sens of Spark Dataframe). This dataframe is made available to R operations such as filter or groupby functions.

```
rdd<-SparkR:::textFile(sc,"MySpark/WaterSpark/T.txt")

lines <- SparkR:::flatMap(rdd,
                function(line) {
                    list(strsplit(line, split = ",")[[1]])
                })

dfR2<-SparkR:::createDataFrame(sqlContext, lines)
df22r <- filter(dfR2, dfR2$D == "22")
```

As seen form the code above, SparkR allows to express analytical queries exactly as if user if executing an R command. Thus, an RDD is firstly created by parallelizing water metering logger file, then by using flatMap and createDataFrame functions, a new dataframe is made available to user for executing filter query on dataset.

5   Conclusion

Smart Water initiatives rely heavily on the use of new technologies such as sensors and smart meters that tend to produce large and uneasily managed volume of data. To manage such database, and beyond the traditional database approach, we presented a novel approach for processing Water Smart Metering datasets using Apache Spark as back end for processing and analyzing large volume of datasets produced by smart meters. As an early conclusion, and with respect to the used version of Spark (1.6), it can be stated that the use of DataFrame seems to be a very efficient, optimized and easy to use method, especially compared to RDD abstractions, for handling Smart Metering Data, and for achieving analytical tasks on them.

References

1. Akyildiz, L.F., Su, W., Sankarasubramaniam, Y., Cayirci, E.: A survey on sensor networks (2002)
2. Bennett, N.D., Croke, B.F.W., Guariso, G., Guillaume, J.H.A., Hamilton, S.H., Jakeman, A. J., Marsili-Libelli, S., Newham, L.T.H., Norton, J.P., Perrin, C., Pierce, S.A., Robson, B., Seppelt, R., Voinov, A.A., Fath, B.D., Andreassian, V.: Position paper : characterising performance of environmental models. Environ. Model. Softw. **40**, 1–20 (2013)
3. Bernardo, V., Curado, M., Staub, T., Braun, T.: Towards energy consumption measurement in a cloud computing wireless testbed. In: Proceedings of the 2011 First International Symposium on Network Cloud Computing and Applications, NCCA 2011, Washington, DC, pp. 91–98. IEEE Computer Society (2011)
4. D'Agostino, D., Clematis, A., Galizia, A., Quarati, A., Danovaro, E., Roverelli, L., Zereik, G., Kranzlmüller, D., Schiffers, M., Felde, N.G., Straube, C., Caumont, O., Richard, E., Garrote, L., Harpham, Q., Jagers, H.R.A., Dimitrijevic, V., Dekic, L., Fiorii, E., Delogu, F., Parodi, A.: The DRIHM project: a flexible approach to integrate HPC, grid and cloud resources for hydro-meteorological research. In: Proceeding of the International Conference for High Performance Computing, Networking, Storage and Analysis, SC 2014, Piscataway, pp. 536–546. IEEE Press (2014)

5. Dunning, T., Friedman, E.: Time Series Databases. O'Reilly Media, Greenwich (2014)
6. Eichinger, F., Pathmaperuma, D., Vogt, H., Muller, E.: Data analysis challenges in the future energy domain. In: Yu, T., Chawla, N., Simoff, S. (eds.) Computational Intelligent Data Analysis for Sustainable Development; Data Mining and Knowledge Discovery Series. CRC Press, Taylor Francis Group, Boca Raton. Chapter 7
7. Vatsavai, R.R., Ganguly, A., Chandola, V., Stefanidis, A., Klasky, S., Shekhar, S.: Spatiotemporal data mining in the era of big spatial data: algorithms and applications. In: Proceedings of the 1st ACM SIGSPATIAL International Workshop on Analytics for Big Geospatial Data, BigSpatial 2012, New York, pp. 1–10. ACM (2012)
8. Fang, X., Misra, S., Xue, G., Yang, D.: Smart grid - the new and improved power grid: a survey. IEEE Commun. Surv. Tutor. (2011)
9. Yigit, M., Cagri Gungor, V., Baktir, S.: Cloud computing for smart grid applications. Comput. Netw. **70**, 312–329 (2014)
10. Zaharia, M., Chowdhury, M., Das, T., Dave, A., Ma, J., McCauley, M., Franklin, M.J., Shenker, S., Stoica, I.: Resilient distributed datasets: a fault-tolerant abstraction for in-memory cluster computing. In: Proceedings of the 9th USENIX Conference on Networked Systems Design and Implementation, NSDI 2012, Berkeley, p. 2. USENIX Association (2012)
11. Zaharia, M., Chowdhury, M., Franklin, M.J., Shenker, S., Stoica, I.: Spark: cluster computing with working sets. In: Proceedings of the 2nd USENIX Conference on Hot Topics in Cloud Computing, HotCloud 2010, Berkeley, p. 10. USENIX Association (2010)
12. Zaharia, M., Das, T., Li, H., Hunter, T., Shenker, S., Stoica, I.: Discretized streams: fault-tolerant streaming computation at scale. In: Proceedings of the Twenty-Fourth ACM Symposium on Operating Systems Principles, SOSP 2013, New York, pp. 423–438. ACM (2013)
13. Laney, D.: META Group, 3D Data Management: Controlling Data Volume, Velocity, and Variety, February 2001
14. Eichinger, F., Pathmaperuma, D., Vogt, H., Müller, E.: Data analysis challenges in the future energy domain. In: Yu, T., Chawla, N., Simoff, S. (eds.) Computational Intelligent Data Analysis for Sustainable Development. Chapman and Hall/CRC, London (2013)
15. http://camel.apache.org/
16. http://sqoop.apache.org/
17. https://kafka.apache.org/
18. http://cassandra.apache.org/

A Cellular Automata Model of Spatio-Temporal Distribution of Species

João Bioco[1]([⊠]), João Silva[1], Fernando Canovas[2], and Paulo Fazendeiro[1]

[1] Instituto de Telecomunicacoes, University of Beira Interior, Covilha, Portugal
`joao.bioco@ubi.pt`
[2] Centro de Ciencias do Mar, University of Algarve, Algarve, Portugal

Abstract. Cellular automata (CA) are discrete models used in several studies due to the capacity to simulate dynamic systems and analyze their behavior. One of the applications of CA in ecology is in the analysis of the spatial distribution of species, where simulation models are created in order to study the response of ecological systems to different kinds of exogenous or endogenous perturbations. In this study we describe an implementation of a cellular automata model able to incorporate environmental data from different sources. To the user is given the power to produce and analyze different scenarios by combining the available variables at will. We present a case study where, departing from a generalized additive model, a possible explanation is given for the distribution of two haplotypes of honeybees along Iberian Peninsula. The results of our model are compared and discussed at the light of the real data collected on the terrain.

Keywords: Environmental modeling · Cellular automata · Modeling tools · Species distribution models

1 Introduction

Cellular automata (CA) are models of natural computing initiated by von Neumann in late of 1940 for the modeling of biological self-reproduction [9]. A cellular automaton (CA) is a discrete dynamic model comprising a orderly grid of cells [10]. Each cell in a CA is a finite state machine. The next state of the cells depends on the states of their neighboring cells according to a local update rule. The grid can be in any finite number of dimensions.

CA starts by local interactions between cells resulting in a complex dynamic system. A classical example of CA is Conway's Game of Life [8].

CA are studied in several areas such as mathematics, computability theory, theoretical biology, ecology, physics and complexity science [19]. Nowadays CA are still being studied and applied in several application domains, as a consequence has been observed the development of a number of variations of CA [10].

© Springer Nature Switzerland AG 2019
M. Ezziyyani (Ed.): AI2SD 2018, AISC 913, pp. 118–128, 2019.
https://doi.org/10.1007/978-3-030-11881-5_11

In this work we synthesize a software tool, able to model and simulate distribution of species through cellular automata. As a case study we create an evolutionary model of Iberian honeybees by using cellular automata in order to infer past, present and future scenarios of evolution. The remaining of this paper is organized as follows: In the second section some studies related to our work are addressed, i.e. cellular automata applied in ecological and epidemiological systems; In the third section a description of our software tool is presented, followed by the case study about distribution of African and European honeybees. The paper ends with a concluding section.

2   Related Work

Since the initial proposal of Von Neumann a plethora of CA have been used not only to study the behavior of biological and physical systems (e.g. [11,13,17]) but also to elicit solutions to several real-life problems (e.g. [1,4,7,10,13,16,18]). In this section we focus our attention in selected applications of CA in ecology and epidemiology.

2.1   Ecological Systems

Cellular automata are widely used in ecological systems [11,13,18]. Various studies report the application of CA in modeling ecological problems, such as CA-based models to simulate and predict the spatial distribution of species [17]; CA-based models to predict the presence of invasive species [12], cellular automata model for land use change [11], etc.

Keshtkar and Voigt [6] proposed a hybrid model (CA-Markov model) composed by cellular automata and Markov chains with the models of species distribution to investigate species migration in the future. They implement a GIS-based cellular automata model, called MigClim [15] to project future distribution over the 21st century for three plant species. A limitation of the study, related to the data, was the absence of independent data available to evaluate the predictive power of the SDM.

Another study that applies cellular automata in ecology is the study of Palmate et al. [17] that applied spatiotemporal land use/land cover (LU/LC) modeling approach to address land resources problems namely: agriculture and water resources management issues. Palmate et al. [17] developed an integrated cellular automata (CA) - Markov Chain (MC) model in order to predict interactively future land use/land cover (LU/LC) scenarios by giving some solutions to current problems related to land resources. The study area was Betwa River Basin (BRB) of Central Índia. According to the results, in this study, overall classification accuracy varies from 77% to 87% and Kappa coefficient varies from 0.709 to 0.836. CA-MC model was capable to successfully discover future problems related to food security and surface water resources availability.

Barbosa et al. [21] implemented a cellular automata model to predict the spread of golden mussel in Brazil, on temporal and spatial scale. Their model assumed cells with a resolution of 4.1 km for the entire Brazilian territory, and they used in their model three fundamental parameters: (a) altitude, river size and presence of waterways, and (b) predicted population density for Brazil in 2015. This model was able to predict satisfactorily the risk of invasion by golden mussel, and the simulation of two specific years (2030 and 2050) showed a high risk of invasion in north and northeastern Brazil.

Mahmoud and Chulahwat [22] proposed a flexible fire propagation model (simulation-based model) by using cellular automata, applying a specific set of rules to model propagation of wildfires. This model considers a set of key parameters such as humidity, nature of vegetation and topology. This simulation-based model shows a good flexibility, allowing adjustments in its accuracy to a certain extent by optimizing the propagation rules using real-event data.

Guimapia et al. [23] proposed a cellular automata-based model to predict the risk of invasion and spread of an invasive insect called T. absoluta, originated from South America, across Africa. Four key factors such as land vegetation cover, temperature, relative humidity and yield of tomato production are used in order to simulate the spreading behavior of the pest. By the simulation is possible to understand the role that each of these keys factor play is the process of propagation of invasion. The study was conducted in order to inform ecologists on the risks that T. absoluta may cause from local to global scales.

2.2 Epidemiological Systems

Epidemiology is another area where the adoption of cellular automata is frequent. Several studies in epidemiology have been using CA to simulate for example, the dissemination of certain disease or to simulate the transmission dynamics of infectious diseases, for prevention and eradication purposes.

Cisse et al. [2] investigated the influence of landscape heterogeneity and host diversity on pathogen transmission in order to understand the dynamics of Chagas disease. They developed an epidemiological model based on the cellular automata, to simulate the spread of Chagas disease in homogeneous and heterogeneous environments with to type of hosts species (competent and non-competent hosts species).

Holko et al. [20] present a new framework based on a two dimensional cellular automata for simulation of the spread of infectious disease in a region in Poland with non-homogeneous spatial population distribution. The model proposed combine solutions such as SEIR (Susceptible, Exposed, Infective and Recovered) model and individual-based model (IBM), with the introduction of several improvements, e.g. real density of the population, parameters reflected the dynamics of population and different types of infection in one numerical system.

Pereira and Schimit [24] implemented a model by cellular automata to study the dengue spreading in a population, taking into account two lattices to model the human-mosquito interaction: (a) lattice for human individuals, and (b) lattice for mosquitos, in order to verify different dynamics in populations. The disease has three different serotypes coexisting in population. Normally, many regions exhibit the presence of only one serotype. Pereira and Schimit developed a framework capable of study the occurrence of two and three serotypes at same time in a population. The model is flexible enough allowing its use to other mosquito-borne disease such us schikungunya, yellow fever and malaria.

3   A CA Model for Quasi-Independent Species

In order to build and simulate our model, we developed a software tool able to deal with a variety problems. Our software tool allows researchers to program and simulate the logic of cellular automaton without a single line of code. Our software tool provides a friendly graphical interface where modelers just drag and drop and connect elements. The software created is specialized for problems where the relation between the environment variables and the welfare of certain group is known. In addition, is also a powerful computation and mathematical tool. Our software was developed using C as the programming language, OpenCL for computing through GPU, SDL2 for multi-threading and querying the hardware specifications, SDL2-image used to load images, GTK+ for rendering the GUI and tinyexpr to solve string encoded mathematical expressions used in the Expression node.

3.1   Nodes and Fields

The main elements of our software are the nodes and fields. The node is the most important element of the software, taking in account that the nodes define every simulations logic. A node is an object that contains a solve function, input and output fields stored in a hash table, that are linked to other nodes' fields.

There are three types of nodes: math, automaton and grid. Math nodes are used to hold numbers and solve all sorts of math operation; Grid nodes are responsible for holding and modifying environment variables as well as RGB images and other bi-dimensional data structures; Automaton nodes are specialized in solving cellular automata.

3.2   B. Saving and Loading the Model

The software tool allows user to save a xml file that contains all model components (nodes and their links). To load the .xml saved file, the engine uses a library called libxml2 that is bundled with gtk+. It automatically loads any given xml file into an xmlNode structure that is then parsed by the program to re-allocate all of the nodes used previously.

Another important particularity of this software is the possibility to save an image of the current state of the simulation at any time during the simulation. It means that we can have several snapshots of the model simulation allowing a comparison of these several states of the simulation with the real data. We believe that this software is powerful enough to be widely used by researchers interesting in modeling and simulating species distribution models using cellular automata.

4   Case-Study: Apis Mellifera - Distribution of African and European Lineages

In our case study we intend to create an evolutionary model of the Iberian honeybee by using cellular automata. The CA model was created in our software tool (engine). Model was simulated using past and present climatic data (climatic variables), to allow the model (ecological system) changing throughout space and time as climate variables changed. Several snapshots of simulation were saved while simulation still running. We run our model until the time that there is no significant alteration (stabilization of the simulation) and then we stopped the simulation. We perform several simulations in order to analyze the differences between the simulation results. After that we perform the comparison between the simulation results with the real results collected in the study area (Iberian Peninsula), in order to validate and evaluate the model.

4.1   Study Area

The study area is located in Iberian Peninsula and covers $582\,000\,\text{km}^2$, with an higher elevation of $3{,}478\,\text{m}$ ($11{,}411\,\text{ft}$) in Mulhacén Mountain, located in province of Granada, Spain, see Fig. 1.

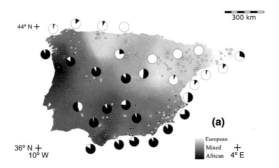

Fig. 1. Localities studied (dots) and the proportion of haplotypes belonging to each evolutionary lineage (fill color for dots following the grey scale bar). Pie charts show the proportion of haplotypes belonging to each evolutionary lineage by Province (black for African and white for European).

4.2   Environmental Data

Environmental data were used to describe every $30 \times 30\,\text{s}$ unit grid of the study area (World Geodetic System 1984 WGS84, geographical coordinates). Estimates of temperature and water availability were obtained from the climatic atlas of the World [14]. For temperature environmental descriptors, the following five variables were used: elevation, annual mean temperature, maximum temperature of the hottest month, minimum temperature of the coldest month and precipitation seasonality index. By seasonality of precipitation was meant the tendency for a place to have more rainfall in certain months or seasons than in others [3].

4.3   The Model

Our honeybee evolutionary model by cellular automata was constructed with a set of rules, based on the influence of the four environmental variables (rfseas, tann, mntcm and mxtwm) in distribution of honeybees. The model aims to find suitable places for a specific type of honeybees (African and European).

An extensive survey of mitochondrial haplotypes in honeybee Apis mellifera colonies from the Iberian Peninsula has corroborated previous hypotheses about the existence of a joint clinal variation of African and west European evolutionary lineages [5]. It has been found that the Iberian Peninsula is the European region with the highest haplotype diversity. The frequency of African haplotypes decreases in a SW-NE trend, while that of European haplotypes increases. So our goal is to apply all the constraints at our disposal, represent the proposed Generalized Additive Model (GAM) within the engine created for this purpose and verify if the results of the simulation are consentaneous (or not) with the real data collected in the field.

Figure 2 shows how the model was constructed in the software tool. In part A the two types of species (African and European) and its parameters are defined; in part B are defined the environmental variables that influence the behavior of each specie; in part C the functions that defines the degree of suitability of the species in a certain location are defined; in part D defines a possible representation of the GAM; in part E the rules of spatial expansion of each specie are defined.

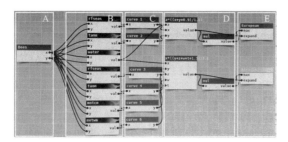

Fig. 2. The model of the case study.

4.4   Results

Several simulations were performed in our software tool in order to gather results regarding to distribution of honeybees in Iberian Peninsula. The initial parameters were set as follows: the maximum number of apiaries of each specie was 500, following a uniform distribution; for each type of species was fixed a spawn rate of 1000, meaning that the number of apiaries will never exceed 1000.

During the simulation these two types of honeybees are moving in order to find the most suitable locations according to the set of rules defined in the model.

In Fig. 3 two types of honeybees can be observed: the color light gray represents European honeybees, and Africans are represented by the color dark gray.

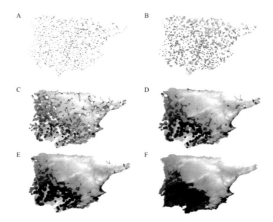

Fig. 3. Six states of the simulation until reach a stabilization point.

Through these simulation states is possible to verify how the two types of honeybees (Europeans and Africans) move along the Iberian Peninsula. The state F of Fig. 3 represents the stabilization of the simulation, where the simulation reaches a stationary point (no visible changes in simulation), after 12100 iterations (epochs). In order to compare simulation results with real data obtained in the study area, two methods were applied: (1) visual comparison and (2) cell-by-cell method. Real data obtained in study area is concerned of the quantity of African and European apiaries in 214 locations (coordinates) in Iberian Peninsula. Five important variables in the dataset were taken into account to do the comparison: longitude, latitude, N (total number of apiaries), the quantity of African honeybees and the quantity of European honeybees. In Fig. 4 is presented two maps; the map in left side shows how the apiaries are distributed in reality based on real data obtained in terrain; the map in right side represents the map resulting of the simulation. Observing these two maps (Fig. 4) i.e. the way that apiaries of real map and the map resulting from simulation are distributed, is unquestionable the high level of similarities between these two maps.

Fig. 4. Real map of Iberian Peninsula with coordinates placed and map obtained from simulation with mismatches points marked.

Therefore, visual comparison is not enough to conclude that a model is performing well because it depends a lot on the level of concentration and also on the quality of the observer's eyes. Because of that, a cell-by-cell comparison is also necessary. Cell-by-cell method consisted in verify the similarities between simulation results and real data, regarding to the quantity of the two types (Europeans and African) in a specific point (coordinate).

Data were distributed in three classes (class A, class E and class AE). Class A represents higher incidence of African honeybees, class E represents higher incidence of European honeybees and class AE represents a mix between African and European honeybees. Setting our model as a classification problem allows a good cell-by-cell comparison, once that real data are also distributed in classes. The dataset with the real data are also distributed in three classes: when there are only African honeybees in a location, is assigned as belonging to class A; when there are only European honeybees in a location, is assigned as belonging to Class E; and where there are the two types of honeybees is assigned as belonging to Class AE. In order to obtain a classification that better approximate to real data, different threshold values (from 0.05 to 0.95, with a step of 0.05) were adopted in classification rule. The rule consists in verify the relative frequency of the two types (A and E) in each point and then assign to a class. Another way to evaluate our model is by calculating a set of common performance metrics normally used in species distribution models.

Table 1. Performance measures of the model: TSS = True Skill Statistic; FPR = False Positive Rate; TPR = True Positive Rate (Threshold = 0.40).

Performance metric	Value
Overall accuracy	0.75
Sensitivity (recall)	0.67
Specificity	0.78
Precision	0.53
F measure	1.02
Kappa	0.42
FPR	0.22
TPR	0.67
TSS	0.45

4.5   Discussion

According to the results, our software engine can be used to simulate distribution of honeybees in Iberian Peninsula. It is possible to verify how our software engine performs the simulation until stabilization point as can be seen in Fig. 3. One important parameter in our software engine is the number of elements, both African and European honeybee defined to start the simulation. These African and European honeybees are generated randomly in a location of the study area. Our simulations were initialized with 1000 as the maximum number of apiaries (both African and European honeybees).

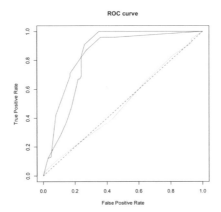

Fig. 5. Roc curve of the model: blue color for African; red color for European and green color for both.

If we observe the map resulting of the simulation (Fig. 3) and the real map of Iberian Peninsula with coordinates placed (Fig. 4) it is possible to verify a great similarity between both maps. In Fig. 4 (left side) the blue points represent locations where the incidence of African honeybees is higher, and the pink points represent locations where the incidence of European honeybees is higher; green points represents locations where exist both African and European. On the other hand in Fig. 3 the dark gray color represents African honeybees and the light gray color represents European honeybees. Using visual comparison is undeniable to verify the similarities between the two maps.

In Fig. 4 (right side) the regions where the results of simulations are different from the real data are marked with a red circle. With careful observation it is possible to verify that mismatches between real data and simulation data are present with higher incidence in locations where we observe the two types of species.

In Table 1 the metrics such as overall accuracy, TSS, kappa, specificity and sensitivity, frequently used to measure the performance of species distribution models are presented. Looking into the performance metrics is possible to verify that our model is performing well. These metrics values were obtained setting a threshold of 0.40 by observing the best results with this limit (overall accuracy: 0.75).

Curves (Fig. 5) were built using different thresholds (from 0.05 to 0.95, with a step of 0.05). The curve with the color blue represents African honeybees, color red represents European honeybees and color green represents locations where the two types of species are observed. According to the graphics, results from simulation of African and European honeybees are performing well if setting a threshold from 0.35 to 0.55. Otherwise, our model present problems in locations where there are both types of species at same time (class AE of our classification model). Figure 4 (right side) confirms this statement, where the locations in frontier are marked with a red circle.

Our model classifies well African honeybees (belonging to class A) and European honeybees (belonging to class E. Locations where exists both types of species (class AE) are difficult to classify.

5   Conclusion

In this study we developed a software tool able to model and simulate distribution of species using cellular automata. We describe an implementation of a cellular automaton model able to incorporate environmental data collected from different heterogeneous sources. Our case study was the distribution of two types of honeybees (African e European) in Iberian Peninsula.

Results obtained by our software are promising. This software tool allows simulating the evolution of ecosystems constituted by non-antagonistic species as well as explaining how this evolution may have happened, by having a mathematical description of how the suitability of a place varies and according to which variables.

In future work we intend to implement another models such as predator-prey, forest models, land use change, population dynamics, etc, using our software tool. Another future work is related to mapping between geological time and computational time during simulations.

References

1. Khan, A.R.: Replacement of some graphics routines with the help of 2D cellular Automata Algorithms for faster graphics operations (2015)
2. Cissé, B., El Yacoubi, S., Gourbiére, S.: A cellular automaton model for the transmission of Chagas disease in heterogeneous landscape and host community. Appl. Math. Model. **40**(2), 782–794 (2016)
3. Markham, C.G.: Seasonality of precipitation in the United States. Ann. Assoc. Am. Geogr. **60**(3), 593–597 (1970)
4. Burks, C., Farmer, D.: Towards modeling DNA sequences as automata. Physica D: Nonlinear Phenomena **10**(1–2), 157–167 (1984)
5. Canovas, F., De la Rúa, P., Serrano, J., Galián, J.: Geographical patterns of mitochondrial DNA variation in Apis mellifera iberiensis (Hymenoptera: Apidae). J. Zool. Syst. Evol. Res. **46**(1), 24–30 (2008)
6. Keshtkar, H., Voigt, W.: Potential impacts of climate and landscape fragmentation changes on plant distributions: coupling multi-temporal satellite imagery with GIS-based cellular automata model. Ecol. Inform. **32**, 145–155 (2016)
7. Régnière, J., Saint-Amant, R., Béchard, A.: BioSim: optimizing pest control efficacy in forestry. Natural Resources Canada, Canadian Forest Service, Laurentian Forestry Centre. Branching out (3), 2 (2003)
8. Conway, J.: The game of life. Sci. Am. **223**(4), 4 (1970)
9. Von Neumann, J., Burks, A.W.: Theory of self-reproducing automata. IEEE Trans. Neural Netw. **5**(1), 3–14 (1966)
10. Bhattacharjee, K., Naskar, N., Roy, S., Das, S.: A survey of cellular automata: types, dynamics, non-uniformity and applications. arXiv preprint arXiv:1607.02291 (2016)

11. Pinto, N., Antunes, A.P., Roca, J.: Applicability and calibration of an irregular cellular automata model for land use change. Comput. Environ. Urban Syst. **65**, 93–102 (2017)
12. Linh, O.T.M., Huong, L.H., Quy, L.T., Huy, N.C., Hiep, H.X.: Simulation the BPH spread with the impact of their natural enemies based on Cellular Automata and Predator-Prey model. In: 2016 Eighth International Conference on Knowledge and Systems Engineering (KSE), pp. 121–126. IEEE (2016)
13. Hewitt, R., Diaz-Pacheco, J.: Stable models for metastable systems? Lessons from sensitivity analysis of a Cellular Automata urban land use model. Comput. Environ. Urban Syst. **62**, 113–124 (2017)
14. Hijmans, R.J., Cameron, S.E., Parra, J.L., Jones, P.G., Jarvis, A.: Very high resolution interpolated climate surfaces for global land areas. Int. J. Climatol. **25**(15), 1965–1978 (2005)
15. Engler, R., Guisan, A.: MigClim: predicting plant distribution and dispersal in a changing climate. Divers. Distrib. **15**(4), 590–601 (2009)
16. Mitra, S., Das, S., Chaudhuri, P.P., Nandi, S.: Architecture of a VLSI chip for modelling amino acid sequence in proteins. In: Proceedings of the Ninth International Conference on VLSI Design, pp. 316-317. IEEE (1996)
17. Palmate, S.S.: Modelling spatiotemporal land dynamics for a trans-boundary river basin using integrated Cellular Automata and Markov Chain approach. Appl. Geogr. **82**, 11–23 (2017)
18. Ghosh, S., Bachhar, T., Maiti, N.S., Mitra, I., Chaudhuri, P.P.: Theory and application of equal length cycle cellular automata (ELCCA) for enzyme classification. In: International Conference on Cellular Automata, pp. 46–57. Springer, Heidelberg (2010)
19. Wolfram, S.: Statistical mechanics of cellular automata. Rev. Mod. Phys. **55**(3), 601 (1983)
20. Holko, A., Mędrek, M., Pastuszak, Z., Phusavat, K.: Epidemiological modeling with a population density map-based cellular automata simulation system. Expert Syst. Appl. **48**, 1–8 (2016)
21. Barbosa, N.P., Ferreira, J.A., Nascimento, C.A., Silva, F.A., Carvalho, V.A., Xavier, E.R., Cardoso, A.V.: Prediction of future risk of invasion by *Limnoperna fortunei* (Dunker, 1857) (Mollusca, Bivalvia, Mytilidae) in Brazil with cellular automata. Ecol. Indic. **92**, 30–39 (2018)
22. Mahmoud, H., Chulahwat, A.: A probabilistic cellular automata framework for assessing the impact of WUI fires on communities. Procedia Eng. **198**, 1111–1122 (2017)
23. Guimapi, R.Y., Mohamed, S.A., Okeyo, G.O., Ndjomatchoua, F.T., Ekesi, S., Tonnang, H.E.: Modeling the risk of invasion and spread of Tuta absoluta in Africa. Ecol. Complex. **28**, 77–93 (2016)
24. Pereira, F.M.M., Schimit, P.H.T.: Dengue fever spreading based on probabilistic cellular automata with two lattices. Phys. A Stat. Mech. Appl. **499**, 75–87 (2018)

A Parallel Approach to Optimize the Supply Chain Management

Otman Abdoun, Yassine Moumen$^{(\boxtimes)}$, and Ali Daanoun

Department of Computer Science, Polydisciplinary Faculty,
Abdelmalek Essaadi University, Larache, Morocco
moumen.yassine@gmail.com

Abstract. The worldwide economic progression in the last century and the Demographic growth has given rise to a huge consumption in the market of goods and services, while globalization decreased the cost of shipping and transportation. The production, transportation, storage and consumption of all these goods, however, have created big environmental problems. Nowadays, global warming, created by large-scale emissions of greenhouse gasses, is a top environmental concern. In this matter, the number of organizations planning to integrate the environmental practices into their future strategic plans is continuously increasing to counter this threat. The environmental benefits of the trend are clear: fewer vehicles burning fuel, crowding urban streets, and taking up valuable parking areas. However, the problem with transportation is that it can be so difficult to choose the perfect path for the vehicle to take if there is many stops to be taking in consideration. Due to the complexity of real world problems, such as supply chain management, finding a good path for vehicles with traditional ways (by using human capabilities) require a long time to satisfy all constraints. Even with machines, this particular problem needs a huge computational power (in term of processing power and memory usage) as well as time to solve. Actually, Parallelism is an approach that not only reduce the resolution time but also improve the quality of the provided solutions. The purpose of this paper is to evaluate the Travelling Salesman Problem (TSP) as a function of forming and optimizing transport networks using an efficient parallelization strategy for the Ant Colony Optimization (ACO) taking the maximum advantage of the parallel architecture offered by NVidia's Graphics Processing Units (GPUs).

Keywords: GPU · Parallel Ant Colony Optimization
Sequential Ant Colony Optimization · Travelling Salesman Problem · CUDA

1 Introduction

Transport has always been the cornerstone of logistical problems. Since its appearance during the Napoleonic campaigns, all the armies have their own logistics office to ensure the supply of food, materials and food. In the industry, especially the transport of goods and employees transport, the logistical aspect has become an essential component of the good functioning of companies. The very competitive aspect makes the search for efficient logistic solutions crucial, reducing the cost of transport,

M. Ezziyyani (Ed.): AI2SD 2018, AISC 913, pp. 129–146, 2019.
https://doi.org/10.1007/978-3-030-11881-5_12

handling, delays in deliveries, for example. This search for efficient solutions is in line with current concerns about reducing CO2 emissions. The problem we are dealing with in this thesis concerns the assignment and routing of a set of homogeneous or heterogeneous vehicles to serve transport demands. The simplest case is the static assignment of a vehicle to a route of stations. This problem of developing vehicle tours known as "Travelling Salesman Problem" (TSP) [1]. It represents a multicriteria combinatorial optimization problem. It belongs to the class of NP-hard problems of combinatorial optimization. In its basic version, the TSP models a transportation problem that involves delivering and collecting Objects (or people) from a set of places of departure to destinations. The solution to this problem consists in determining a set of rounds by minimizing several criteria such as the total distance travelled, the amount of used resources, etc. One of the hardest known NP-Hard Problem is TSP (Traveling Salesman Problem), which take an enormous amount of time to calculate a solution [2].

In this context, some metaheuristic methods has been developed based on natural phenomenon observation, for example Genetic Algorithm (AG), Ant Colony Optimization (ACO), simulated annealing and Hybridation GA & ACO [3]. This method are suitable to solve NP-hard optimization problems in moderate execution times. However, yet none of these algorithms have been able to reach the optimal solution for large-scale problem instances, and since there is no exact algorithm to solve an optimization problem in polynomial time, the minimal expected time to obtain optimal solution is exponential. Therefore, it is only possible to use metaheuristic algorithms to find approximate solutions for a given problem (a "good" solution).

Some of the algorithms can be implemented in a parallel environment to improve the performance. The popular method to parallelize an algorithm is to have a cluster of computers each with its own processor and memory, split the problem into sub-problems, and then assign each one to a computer. But then again, by adding computational power into the cluster, we can face the problem of overloading the network connecting the computers, which will increase the running time for the algorithm. However, we can overcome this problem by using other methods for example: OpenMP [4], GPU [1, 5], MPI, etc.

We have come up with a new approach that exploit the full potential power of the Graphical processing units (GPUs) are which are a specialized processors with dedicated memory that conventionally perform floating point operations required for rendering graphics. In response to commercial demand for real-time graphics rendering, the current generation of GPUs have evolved into many-core processors that are specifically designed to perform data-parallel computation. Due to the inherently parallel nature of Genetic Algorithm, it is relatively easy way to implement on GPUs. However, it also brings some significant challenges due to its synchronization points and memory access patterns.

This paper is organized as following: first addressing the problematic. The basic principles of GPU computing are outlined in Sect. 2. An introduction to Traveling Sales Man Problem in Sect. 3. A brief description of the Ant Colony Optimization considered in this study is given in Sect. 4. A parallel design and implementation on the GPUs is provided in Sect. 5. Related work in Sect. 6. Finally, concluding remarks are drawn in Sect. 6.

2   GPU Computing

In recent years, GPUs have become increasingly attractive for general-purpose parallel computation. A good use of parallel programming on GPU hardware may produce results surpass tens of traditional CPUs, at a fraction of the cost. In addition, the ever-increasing requirements of the video-game market make the GPU rate of growth much higher than that of traditional microprocessors. Moreover, CPU hardware is suitable for generic sequential codes, being mostly dedicated to non-computational tasks, such as branching. Instead, GPU is design to the optimization of graphics operations, which are parallel by nature. As a result, GPUs dedicate a greater number of transistors to computation, thus achieving an enormous arithmetical intensity.

Compute-oriented APIs expose the massively parallel architecture of the GPU to the developer in a C-like programming paradigm. These commonly used APIs include NVIDIA CUDA, Microsoft DirectCompute, and OpenCL. In this paper we have used CUDA, currently the most popular GPU computing API, but the concepts it presents are readily applied to other APIs. It allows the programmer to access computing resources on the GPU such as video memory, shading units, and texture units directly, without having to program the graphics pipeline. From a hardware perspective, a CUDA-enabled graphics card comprises SGRAM memory and the GPU chip (a collection of streaming multiprocessors or MPs) and on-chip memory.

CUDA includes C/C++ software-development tools, function libraries, and a hardware-abstraction mechanism that hides the GPU hardware from developers. Even though CUDA has greatly simplified the implementation of GPU-enabled applications, a number of hardware and software constraints must still be addressed in order to achieve high performance. In order to understand how to deal with such limitations, the comprehension of some simple basic concepts of CUDA is fundamental. According to CUDA, data-parallel portions of an application are implemented as a multithreaded program called a kernel. The main CPU, acting as the host, can initiate one kernel at a time. Each kernel is executed in parallel by several threads. The programmer groups threads into blocks, which are logically aggregated into a grid (Fig. 1). When a CUDA application is executed, threads are scheduled in groups of warps. A warp executes one common instruction at a time, so full efficiency is realized when all threads of a warp share the same execution path.

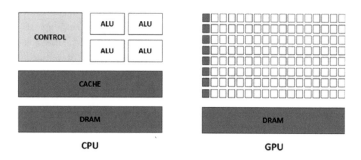

Fig. 1. GPU architecture

The organization of the GPUs memory mirrors the hierarchy of the threads (Fig. 2). Global memory is randomly accessible for reading and writing by all threads in the application. Shared memory provides storage reserved for the members of a thread block. Local memory is allocated to threads for storing their private data. Last, private registers are divided among all the threads residing on the MP. Registers and shared memory, located on the GPU chip, have much lower latency than local and global memories, implemented in SGRAM. However, global memory can store several gigabytes of data, far more than shared memory or registers. Furthermore, while shared memory and registers only hold data temporarily, data stored in global memory persists beyond the lifetime of the kernels (Fig. 3).

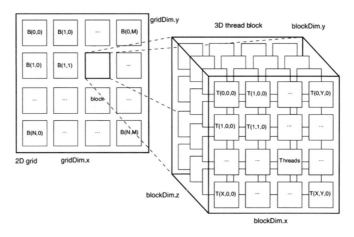

Fig. 2. CUDA program structure and memory hierarchy

The host and the device (i.e., the GPU) communicate by copying data from/to the CPU's memory to/from the so-called GPU global memory. Therefore, by default, threads operate on data stored in the GPU's global memory. This kind of memory is off-chip, and features huge latency times. Other memory devices - faster than global memory - are available, and CUDA provides the software construct to transfer data to/from the global memory from/to these devices.

It is a programmer's responsibility to carefully design and implement kernel codes in order to minimize the number of global-memory reads, by making use of the other available kinds of memory. Moreover, when global memory must be used, it is essential that threads access consecutive array elements, so that memory reads are automatically combined (coalesced) into larger reads, thus reducing the effective number of memory transactions.

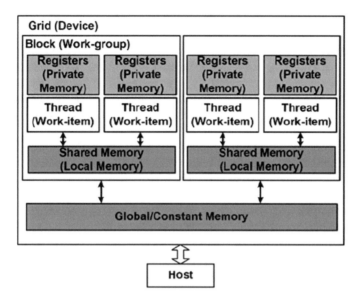

Fig. 3. CUDA memory architecture

3   Problem Addressed

Logistics have distinct geographical dimension, expressed in terms of flows, nodes and networks within the supply chain. If we observe distribution of goods and transportation, it can be considered that its efficiency is proportional to degree of the transport network construction, and strategic distribution planning should be based on the optimal movement drove of a transport network. Transport network flexibility and adaptability can be achieved by optimizing routes for vehicles moving from one place of departure to specific, more than one transport destination. The most important operational decision related to transportation in a supply chain is routing and planning. Most of the transportation problems can be reduce to a simpler representation of Traveling Salesman Problem (TSP).

The Viennese mathematician Karl Menger made the first statement of the Traveling Salesman Problem (TSP) as we know it today in 1930. It arose in connection with "A new definition of curve length" that Menger proposed. As he defined the length of a curve as the least upper bound of the set of all numbers that could be obtained by taking each finite set of points of the curve and determining the length of the shortest polygonal graph joining all the points. "We call this the messenger problem, because in practice the problem has to be solved by every postman, and also by many travellers: finding the shortest path joining all of a finite set of points whose distances from each other are given.

Of course, the problem can be solved by a finite number of trials. However, there is no such a rule that would reduce the number of trials to less than the number of permutations of the given points. The rule of proceeding from the origin to the nearest point, then to the nearest point to that, and so on, does not generally give the shortest path" [1].

The TSP is stated as, given a complete graph, G, with a set of vertices, V, a set of edges, E, and a cost, Cij, associated with each edge in E. The value Cij is the cost incurred when traversing from vertex $i \in V$ to vertex $j \in V$. Given this information, a solution to the TSP must return the shortest Hamiltonian cycle of G (A Hamiltonian cycle is a cycle that visits each node in a graph exactly once. This is referred to as a tour in TSP terms).

Traveling Salesman Problem is one of the most studied combinatorial problems because it is simple to comprehend but hard to solve [2]. The problem is to find the shortest tour of a given number of cities which visits each city exactly once and returns to the starting city [3]

At the first sight, TSP seems to be limited for a few application areas; however, it can be used to solve tremendous number of problems. Some of the application areas are; printed circuit manufacturing, industrial robotics, time and job scheduling of the machines, logistic or holiday routing, specifying package transfer route in computer networks, and airport flight scheduling.

As one of the best-known NP-hard problems, it means that there is no exact algorithm to solve it in polynomial time. The minimal expected time to obtain optimal solution is exponential. Therefore, we usually use heuristics to help us to obtain a "good" solution. Many algorithms were applied to solve TSP with more or less success. There are various ways to classify algorithms, each with its own merits. The basic characteristic is the ability to reach optimal solution: exact algorithms or heuristics.

There are various approaches to solve TSP the classical approaches are dynamic programming, branch and bound which uses heuristic and exact method and results into exact solution. Still, TSP is an NP-hard problem so the time complexity of these algorithms are exponential. Therefore, they can solve the small problem in optimal time but as compared to the large problem time taken by these algorithms are quite high. Unfortunately no classical approach can solve this type of problem in reasonable time as the size of the problem increases complexity increases exponentially.

Many alternate approaches are used to solve TSP, which may not give the exact solution but an optimal solution in a reasonable time. Methods like nearest neighbour, spanning tree based on the greedy approach are efficiently used to solve such type of problems with small size. To overcome this different other approaches based on natural and population techniques such as genetic algorithm, stimulated annealing, bee colony optimization, particle swarm optimization etc. are inspired from these techniques.

However, those methods always fail to find an acceptable solution in a reasonable time. Which leaves us with two options; either minimize the size of the problem or give up and accept the poor results. Luckily, some of the algorithm such as Ant Colony Optimization (ACO) can be implemented in a parallel environment to improve the performance.

4 Ant Colony Optimization

4.1 Mode of Operation

ANT colony optimization [4, 5] technique introduced by Marco Dorigo in 1991 is based upon the real ant behavior in finding the shortest path between the nest and the food. They achieved this by indirect communication by a substance called pheromone, which shows the trail of the ant. Ant uses heuristic information of its own knowledge the smell of the food and the decision of the path travelled by the other ants using the pheromone content on the path. The role of the pheromone is to guide other ants towards the food.

Ant has the capability of finding the food from their nest with the shortest path without having any visual clues. At a given point where there are more than one path to reach to their food then ants distribute themselves on different paths and the path and lay pheromone trace on that path and return with same path. Thus the path with minimum distance will acquire more pheromone as compared to other paths as the ants will return faster from that path comparative to the other path. Therefore, the new ants coming in the search of food will move with probability towards the path having higher pheromone content as compared to the path having lower pheromone content and in the end, all the ants will move towards the same path with the minimum shortest path to their food (Fig. 4).

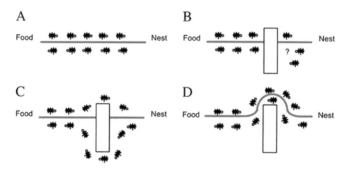

Fig. 4. Real ant behavior in finding the shortest path

4.2 Sequential Ant Colony Optimization

Real ants can indirectly communicate by pheromone information (a chemical substance that attract other ants searching for food). The ant release pheromone on her path while walking, and the other ants follow the pheromone trails with some probability. The more ants walk on a trail, the more pheromone is released on it and more and more ants follow the trail.

The important characteristic of pheromones is evaporation. This process depends on the time. When the ants no longer use the path, pheromones are more evaporated and the ants begin to use other paths.

The ant colony optimization algorithm is made up of only a few steps. First, each ant in the colony constructs a solution based on the existed pheromone trails. Next ants will release pheromone trails on the path, depending on the solution's quality. In the example of the traveling salesman problem, this would be the paths between the cities.

Finally, after all ants have finished constructing a solution and laying their pheromone trails, pheromone is evaporated from each edge depending on the pheromone evaporation rate. These steps are repeated many times to insure the best quality solution as following:

Construct solutions: At the start, each path between cities has some initial amount of pheromone. Each ant starts from a random city and goes from a city to the next until all cities are visited exactly once and then the ant returns to the start city.

Pheromone Update: The ants update pheromones on paths connecting the cities, every time an ant crosses an edge, it add some pheromone to the existing one (Fig. 5).

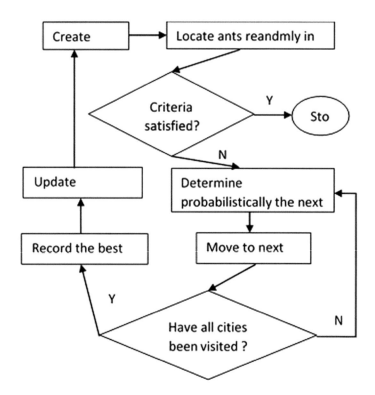

Fig. 5. Sequential Ant Colony Optimization

Evaporation: After all ants complete their trip, evaporation is applied to all paths between cities, which subtract an amount of pheromone.

Steps of the algorithm:

1. Create a colony of artificial ants.
2. Put each ant on a starting city.
3. Move all ants to next city. The ant in city X selects the next city to visit by calculating probabilities (including: partial solution, distance between city X and city Y, amount of pheromone on the path).
4. For every ant Add pheromone to path.
5. Memories the best solution.
6. Subtract pheromone value from all edges.
7. Repeat step 1 to 6 until we get an optimal solution to the problem.

5   Parallel Ant Colony Optimization

Most optimization tasks found in real-world applications impose several constraints that usually do not allow the utilization of exact methods. The complexity of these problems (they are often NP-hard) or the limited resources available to solve them (time, memory) have made the development of metaheuristics a major field in operations research. In these cases, metaheuristics provide optimal or suboptimal feasible solutions in a reasonable time. Although the use of metaheuristics allows to significantly reduce the time of the search process, the high dimension of many tasks will always pose problems and result in time-consuming scenarios for industrial problems. Therefore, parallelism is an approach not only to reduce the resolution time but also to improve the quality of the provided solutions. The latter holds since parallel algorithms usually run a different search model with respect to sequential ones.

The Ant Colony Optimization approaches require a large number of simulations, and to employ this approaches in a real-time manner would require that the solutions could be executed in a more practical run-time. For executing the ACO approach, artificial ants in each trial and their paths are independently created based on pheromone levels. Therefore, the artificial ants can be distributed over to the GPU processors to save time, as we will see next.

Artificial ants are stochastic construction procedures that probabilistically build a solution by iteratively adding solution components to partial ones by taking into account heuristic information on the problem and pheromone trails, which change dynamically at runtime to reflect the acquired search experience.

The concurrent nature of both tour construction and global search of the solution space makes the ACO metaheuristic a good candidate for parallelization. However, this potential comes with important challenges mainly due to pheromone management and to the size of the data structures that have to be maintained. Works on traditional, GPU-based parallel ACO can be classified into two general approaches: ANT global block and ANT shared block.

In the parallel ants general strategy, ants of a single colony are distributed to processing elements in order to execute tour constructions in parallel. On a conventional CPU architecture, the concept of processing element is usually associated to a singlecore processor or to one of the cores of a multi-core processor.

On a GPU architecture, the obvious choice is to associate this concept to a single SP. In that case, a first strategy that may be defined is to associate each ant to a CUDA thread. Each thread then computes the state transition rule of each ant in a SIMD fashion. We call this strategy Parallel Ant Colony Optimization PACO. It has the advantage of allowing the execution of a great number of ants on each SM and the drawback of limiting the use of fast GPU memory. In fact, each ant needs its own data structures, mainly tour and probability arrays (of size $O(n)$), to effectively compute the state transition rule required to build a solution.

Simple calculations show that using the shared memory for these structures would restrict the algorithm to use a very small number of ants on a single SM and that this restriction would grow linearly with problem size. Code optimizations may help raise that number by a constant factor, but hardly enough to bypass algorithmic limitations. Therefore, these data structures must be stored in global memory and accessed in read/write mode during the tour construction phase.

The second proposed strategy is based on associating the concept of processing element to a whole SM. In that case, each ant is associated to a CUDA block and parallelism is preserved for tour construction. We call this strategy Global Parallel Ant Colony Optimization GPACO. A single thread of a given block is still in charge of executing the tour construction of an ant, but an additional level of parallelism may be exploited in the computation of the state transition rule.

In fact, an ant evaluates several candidate cities before selecting the one to add to its current solution. As these evaluations can be done in parallel, they are assigned to the remaining threads of the block. Following the idea of the first strategy, a simple implementation would then imply keeping ant's private data structures in the global memory. However, as only one ant is assigned to a block and so to a SM, taking advantage of the shared-memory becomes possible for problems bigger than a few dozen cities. Data needed to compute the ant state transition rule is then stored in this memory that is faster and accessible by all threads that participate in the computation. In order to evaluate the benefits and limits of using the shared-memory in this context, two variants of the GPACO strategy are distinguished: ANT global block and ANT shared block.

Most remaining issues encountered in the GPU implementation of the parallel ants general strategy are related to memory management. More particularly, data transfers between CPU and GPU as well as global memory accesses require considerable time. These accesses may be reduced by storing the related data structures in shared memory. However, in the case of ACO, the three central data structures are the pheromone matrix, the distance matrix and the candidates lists, which are needed by all ants of the colony while being too large (ranging from $O(n * cl)$ to $O(n2)$ in size) to fit in shared memory. They are then kept in global memory.

On the other hand, as they are not modified during the tour construction phase, it is possible to take benefit of the texture cache to reduce their access times. Also, in order to compute the state transition rule, random numbers need to be generated and that feature is not directly available on GPUs. For that matter, the adopted solution is to compute them prior to the beginning of the iterations and to store them in texture memory to enable faster access (Fig. 6).

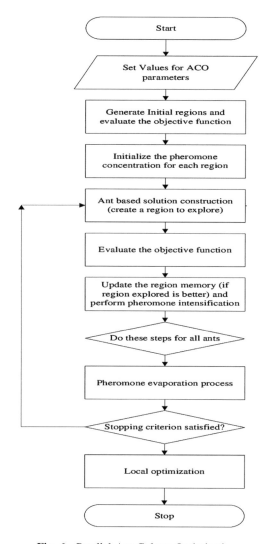

Fig. 6. Parallel Ant Colony Optimization

In general, a parallel ant colony optimization Algorithm follows the same pattern:

1. Create a colony of artificial ants.
2. Associate each ant to thread (or a block).
3. Put each ant on a starting city.
4. Move all ants to next city at once.
5. Update pheromone locally for the ant.
6. Memories the best solution.
7. Update pheromone globally for all ants.
8. Subtract pheromone value from all edges.
9. Repeat step 1 to 8 until we get an optimal solution to the problem.

6 Related Work

Parallel implementations became popular in the last decade in order to improve the efficiency of population-based metaheuristics. By splitting the population into several processing elements, parallel implementations of metaheuristics allow reaching high quality results in a reasonable execution time, even when facing hard to solve optimization problems [2]. Parallel algorithms not only take benefit of using several computing elements to speed up the search, they also introduce a new exploration pattern that is often useful to improve over the result quality of the sequential implementations.

Many papers can be found in the related literature stating that parallel implementations are useful to improve the ACO exploration pattern; However, researchers often lack a generalized point of view, since they usually tackle a unique implementation to solve a specific problem.

Stützle, Hoos and Talbi et al. [39, 40] first suggested the application of parallel computing techniques to enhance both the ACO search and its computational efficiency, while Randall and Lewis [84] proposed the first classification of ACO parallelization strategies. The book chapter by Janson et al. [57] and the article by Ellabib et al. [46] are the only previous works that have collected bibliography of published papers proposing parallel ACO implementations. Janson et al. reviewed parallel ACO proposals published up to 2002, focusing on comparing "parallelized" standard ACO algorithms, specific parallel.

ACO methods, and hardware parallelization; although they did not include an explicit algorithmic taxonomy. Ellabib et al. briefly commented parallel ACO implementations up to 2004, focusing in describing the applications, and they only distinguished between coarse-grain and fine-grain models for parallel ACO.

The classic proposals of parallel ACOs focused on traditional supercomputers and clusters of workstations. Nowadays, the novel emergent parallel computing architectures such as multicore processors, graphics processing units (GPUs), and grid environments provide new opportunities to apply parallel computing techniques to improve the ACO search results and to lower the required computation times.

7 Experimental Results

For each general parallelization approach, the two specific GPU strategies are experimented and compared on various TSPs with sizes varying from 51 to 229 cities. Minimums and averages are computed from 10 trials for. An effort is made to keep the algorithm and parameters as close as possible to the Sequential program. In some cases where we propose slightly different parameters to make the algorithm a better fit for GPUs, our choices are justified and results are provided to show the impact on solution quality.

The relative speedup metric is computed on mean execution times to evaluate the performance of the proposed implementations. For the PACO and GPACO strategies involving only one GPU, speedups are calculated by dividing the sequential CPU time with the parallel time, which is obtained with the same CPU and the GPU acting as a co-processor. Experiments were made on one GPU of a NVIDIA GTX 1060. Each GPU contains 14 SMs, 32 SPs per SM, 48 KB of shared memory per SM and a warp size of 32. The machine also includes one 4-core Xeon E3-1220 CPUs running at 3.10 GHz and 8 GB of DDR3 memory. Application code was written in the C++ for CUDA v8 programming environment. For each problem, the number of threads and blocks used for the GPU resolution were empirically chosen according to a preliminary study based on our previous works. At this point, optimal general configurations can hardly be determined beforehand since they depend on many technical constraints linked to the GPU architecture and programming environment as well as on the algorithmic design of the metaheuristic.

Overall, even though it is generally recommended to use a high number of threads in GPU applications, a compromise had to be found in the case of ACO algorithms. Consequently, thread and block configurations are provided for all experiments.

The solution quality as well as the execution time of the ACO mainly depends on the parameter decisions. The parameters of the algorithm chosen in this study are as shown in Table 2.

A first step in our experiments is to compare solution quality obtained by sequential and parallel versions of the algorithm. Table 1 presents minimum and average tour lengths for each strategy and for each problem. The reader may first note the similarity between the results obtained by our sequential and the parallel implementation for the smaller instances. Results provided for all parallel strategies are also similar, showing that solution quality is globally preserved. A second step is to evaluate and compare the reduction of execution time that is obtained with each parallelization strategy.

Table 1. Details of the underlying system

	CPU	GPU
Manufacturer	Intel	NVIDIA
Model	Xeon E3-1220 V2	GTX 1060
Architecture	x86-64	Pascal
Clock frequency	3.10 GHz	1544 MHz
Cores	4	1280
DRAM memory	8 GB	6 GB

Table 2. Parameters of the algorithm

Parameter	Value
Number of cities	51, 100, 150 and 229
Number of ants	10, 100 and 1000
Alpha	2.0
Beta	3.0
ROU	0.5
Iterations	10, 50 and 100

Figure 7 shows the speedups obtained for each problem. The reader may notice PACO speedups is lower than GPACO for each problem and the gap becomes larger as problem size increases. This difference comes mainly from code divergence induced by computing the state transition rule of many ants on the same block in SIMD mode, as well as from the limited amount of threads and blocks required to effectively hide memory latencies. Nevertheless, speedup generally increases with problem size, indicating that the strategy is scalable to some extent.

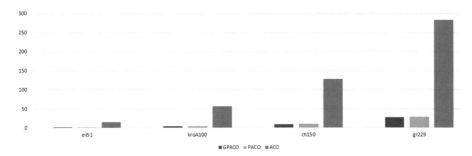

Fig. 7. 1000 ants in 100 iterations

8 Conclusion

The aim of this paper was to propose efficient parallelization strategies for the implementation of Ant Colony Optimization on Graphics Processing Units. Following the parallel ants general approach, the PACO and GPACO strategies aimed at associating the ants tour construction and local search phases to the execution of streaming processors and multiprocessors respectively. We showed that both general approaches can be efficiently implemented on a GPU architecture. In fact, the PACO strategy managed to provide speedups as high as 19.47. Speedups raise even higher with the GPACO strategy, reaching 23.60. Overall, this shows that it is possible to significantly reduce the execution time of ACO on GPU while rigorously keeping the similar competitive solution quality. Still, as it is the case in the field of parallel ACO and parallel metaheuristics in general, much can still be done for the effective use of GPUs. In fact, the variety of the proposed strategies and the extensive comparative study provided in this paper brings its share of questions and research avenues. For example, even though the use of the GPU shared memory leads to some of the best speedups, this hardware feature also shows its limits on bigger TSPs. Moreover, maximal exploitation of GPU resources often requires algorithmic configurations that do not let ACO perform an effective exploration and exploitation of the search space. Globally, this paper shows that parallel performance is strongly influenced by the combined effects of parameters related to the metaheuristic, the GPU technical architecture and the granularity of the parallelization. As it becomes clear that the future of computers no longer relies on increasing the performance on a single computing core but on using many of them in a single system, it becomes desirable to adapt optimization tools for parallel execution on architectures like GPUs. Following this line of thought, our future works are aimed at using the framework and knowledge built in this paper to propose an ACO metaheuristic that is specifically tailored to GPU execution. Also, in order to provide better insight into the memory and algorithmic bottlenecks that have been identified in this work, a more formal analysis would be likely required. For example, the requirements for the different types of memory (main, device, shared, etc.) could be analyzed as a function of problem size, numbers of ants, colonies, SM, SP, etc. Such an analysis could lead to the proposition of algorithms that automatically determine effective thread/block/GPU configurations for ACO and other metaheuristics. We believe that the global acceptance of GPUs as components for optimization systems requires algorithms and software that are not only effective, but also usable by a wide range of academicians and practitioners.

Annex

eil51

Ants	Iterations	GPACO	PACO	ACO
10	10	0.513	0.611	0.017
	50	0.898	1.004	0.092
	100	1.274	1.425	0.175
100	10	0.537	0.664	0.158
	50	0.937	1.045	0.818
	100	1.435	1.544	1.595
1000	10	0.536	0.635	1.449
	50	1.028	1.045	7.417
	100	1.478	1.572	14.97

kroA100

Ants	Iterations	GPACO	PACO	ACO
10	10	0.827	0.91	0.101
	50	2.043	2.132	0.313
	100	3.562	3.743	0.612
100	10	0.907	0.899	0.56
	50	2.191	2.324	2.849
	100	3.946	4.079	5.901
1000	10	0.919	0.941	5.392
	50	2.338	2.424	28.092
	100	4.214	4.378	56.459

ch150

Ants	Iterations	GPACO	PACO	ACO
10	10	1.244	1.281	0.162
	50	4.002	4.021	0.692
	100	7.524	7.604	1.379
100	10	1.464	1.358	1.241
	50	4.357	4.429	6.394
	100	8.204	8.377	12.634
1000	10	1.409	1.518	12.106
	50	4.802	5.007	63.058
	100	9.114	10.012	127.547

gr229

Ants	Iterations	GPACO	PACO	ACO
10	10	3.069	3.112	0.373
	50	11.055	11.032	1.512
	100	21.88	21.303	3.019
100	10	3.246	3.654	2.77
	50	12.981	13.25	14.23
	100	23.58	25.132	28.566
1000	10	3.428	3.412	27.915
	50	14.08	14.965	143.651
	100	27.702	28.371	283.743

References

1. Menger, K.: Ergebnisse eines Kolloquiums **3**, 11–12 (1930)
2. Yang, F.: Solving Traveling Salesman Problem Using Parallel Genetic Algorithm and Simulated Annealing (2010)
3. Philip, A., Taofiki, A.A., Kehinde, O.: A genetic algorithm for solving travelling salesman problem. Int. J. Adv. Comput. Sci. Appl. (2011)
4. Sivanandam, S.N., Deepa, S.N.: Introduction to Genetic Algorithms. Springer, Heidelberg (2008)
5. Konfr˘st, Z.: Parallel genetic algorithms: advances, computing trends, applications and perspectives. In: 18th International Parallel and Distributed Processing (2004)
6. Cant-Paz, E.: Efficient and Accurate Parallel Genetic Algorithms. Kluwer Academic Publishers, New York (2001)
7. Lefohn, A.E., Sengupta, S., Kniss, J.O.E., Strzodka, R., Owens, J.D.: Glift: generic, efficient, random-access GPU data structures. ACM Trans. Graph. **25**(1), 60–99 (2006)
8. Mendez-Lojo, M., Burtscher, M., Pingali, K.: A GPU implementation of inclusion-based points-to analysis. ACM SIGPLAN Notices **47**, 107–116 (2012)

9. Yang, C.-H., Nygard, K.E.: The effects of initial for time constrained population traveling in genetic search salesman problems, pp. 378–383. ACM (1993)

10. Bartal, Y., Gottlieb, L.-A., Krauthgamer, R.: The traveling salesman problem: low-dimensionality implies a polynomial time approximation scheme. In: Proceedings of the 44th Symposium on Theory of Computing - STOC 2012, pp. 663–672 (2012)

11. Fekete, P., Meijer, H., Rohe, A., Tietze, W.: Solving a "Hard" problem to approximate an "Easy" one: heuristics for maximum matchings and maximum traveling salesman problems. In: Algorithm Engineering and Experimentation. Springer, Heidelberg (2001)

12. Alba, E.: Parallel evolutionary algorithms can achieve super-linear performance. Inf. Process. Lett. **82**, 7–13 (2002)

13. Alba, E., Leguizamon, G., Ordonez, G.: Two models of parallel ACO algorithms for the minimum tardy task problem. Int. J. High Perf. Syst. Archit. **1**(1), 50–59 (2007)

14. Bai, H., Yang, D., Li, X., He, L., Yu, H.: Max-min ant system on GPU with CUDA. In: Fourth International Conference on Innovative Computing, Information and Control, pp. 801–804. IEEE (2009)

15. Barr, R.S., Hickman, B.L.: Reporting computational experiments with parallel algorithms: issues, measures and experts' opinions. ORSA J. Comput. **5**(1), 2–18 (1993)

16. Bullnheimer, B., Kotsis, G., Strauss, C.: Parallelization strategies for the ant system. In: De Leone, R., Murli, A., Pardalos, P., Toraldo, G. (eds.) High Performance Algorithms and Software in Nonlinear Optimization. Applied Optimization, vol. 24, pp. 87–100. Kluwer, Dordrecht (1997)

17. Catala, A., Jaen, J., Mocholi, J.: Strategies for accelerating ant colony optimization algorithms on graphical processing units. In: IEEE Congress on Evolutionary Computation, pp. 492–500. IEEE Press (2007)

18. Chu, D., Till, M., Zomaya, A.: Parallel ant colony optimization for 3D protein structure prediction using the HP lattice model. In: 19th IEEE International Parallel and Distributed Processing Symposium, vol. 7. IEEE Computer Society (2005)

19. Craus, M., Rudeanu, L.: Parallel framework for ant-like algorithms. In: Third International Symposium on Parallel and Distributed Computing, ISPDC/HeteroPar 2004, pp. 36–41 (2004)

20. CUDA: Computer Unified Device Architecture Programming Guide 3.1 (2010). http://www.nvidia.com

21. Delisle, P., Gravel, M., Krajecki, M.: Multi-colony parallel ant colony optimization on SMP and multi-core computers. In: Proceedings of the World Congress on Nature and Biologically Inspired Computing, NaBIC 2009, pp. 318–323. IEEE (2009)

22. Delisle, P., Gravel, M., Krajecki, M., Gagné, C., Price, W.L.: A shared memory parallel implementation of ant colony optimization. In: 6th Metaheuristics International Conference, MIC 2005, Vienna, pp. 257–264 (2005)

23. Delisle, P., Gravel, M., Krajecki, M., Gagné, C., Price, W.L.: Comparing parallelization of an ACO: message passing vs. shared-memory. In: Blesa, M.J., Blum, C., Roli, A., Sampels, M. (eds.) LNCS, vol. 3636, pp. 1–11. Springer, Heidelberg (2005)

24. Delisle, P., Krajecki, M., Gravel, M., Gagné, C.: Parallel implementation of an ant colony optimization metaheuristic with OpenMP. In: International Conference on Parallel Architectures and Compilation Techniques, 3rd European Workshop on OpenMP, EWOMP 2001, Barcelona (2001)

25. Delévacq, A., Delisle, P., Gravel, M., Krajecki, M.: Parallel ant colony optimization on graphics processing units. In: Arabnia, H.R., Chiu, S.C., Gravvanis, G.A., Ito, M., Joe, K., Nishikawa, H., Solo, A.M.G. (eds.) Proceedings of the 16th International Conference on Parallel and Distributed Processing Techniques and Applications, PDPTA 2010, pp. 196–202. CSREA Press, Athens (2010)

26. Doerner, K., Hartl, R., Benker, S., Lucka, M.: Parallel cooperative savings based ant colony optimization-multiple search and decomposition approaches. Parallel Process. Lett. **16**(3), 351–370 (2006)

27. Dorigo, M., Gambardella, L.M.: Ant colonies for the traveling salesman problem. BioSystems **43**, 73–81 (1997)

28. Dorigo, M., Stützle, T.: Ant Colony Optimization. MIT Press, Bradford Books, Cambridge (2004)

29. Ellabib, I., Calamai, P., Basir, O.: Exchange strategies for multiple ant colony system. Inf. Sci. **177**(5), 1248–1264 (2007)

30. Islam, M.T., Thulasiraman, P., Thulasiram, R.K.: A parallel ant colony optimization algorithm for all-pair routing in manets. In: 17th International Symposium on Parallel and Distributed Processing. IEEE Computer Society (2003)

31. Li, J., Hu, X., Pang, Z., Qian, K.: A parallel ant colony optimization algorithm based on fine-grained model with GPU-acceleration. Int. J. Innov. Comput. Inf. Control **5**(11(A)), 3707–3716 (2009)

32. Lin, S.: Computer solutions for the traveling salesman problem. Bell Syst. Tech. J. **44**, 2245–2269 (1965)

33. Manfrin, M., Birattari, M., Stützle, T., Dorigo, M.: Parallel ant colony optimization for the traveling salesman problem. In: LNCS, vol. 4150, pp. 224–234 (2006)

34. Middendorf, M., Reischle, F., Schmeck, H.: Multi colony ant algorithms. J. Heuristics **8**(3), 305–320 (2002)

35. Randall, M., Lewis, A.: A parallel implementation of ant colony optimization. J. Parallel Distrib. Comput. **62**(2), 1421–1432 (2002)

36. Scheuermann, B., Janson, S., Middendorf, M.: Hardware-oriented ant colony optimization. J. Syst. Archit. **53**, 386–402 (2007)

37. Scheuermann, B., So, K., Guntsch, M., Middendorf, M., Diessel, O., ElGindy, H., Schmeck, H.: FPGA implementation of population-based ant colony optimization. Appl. Soft Comput. **4**, 303–322 (2004)

38. Stützle, T.: Parallelisation strategies for ant colony optimization. In: Eiben, A., Bäck, T., Schwefel, H.-P., Schoenauer, M. (eds.) Proceedings of the Fifth International Conference on Parallel Problem Solving from Nature. PPSN V. Springer, New York (1998)

39. Stützle, T., Hoos, H.: Max-min ant system. Future Gener. Comput. Syst. **16**(8), 889–914 (2000)

40. Talbi, E., Roux, O., Fonlupt, C., Robillard, D.: Parallel ant colonies for the quadratic assignment problem. Future Gener. Comput. Syst. **17**(4), 441–449 (2001)

41. Wang, J., Dong, J., Zhang, C.: Implementation of ant colony algorithm based on GPU. In: Banissi, E., Sarfraz, M., Zhang, J., Ursyn, A., Jeng, W.C., Bannatyne, M.W., Zhang, J.J., San, L.H., Huang, M.L. (eds.) Sixth International Conference on Computer Graphics, Imaging and Visualization: New Advances and Trends, pp. 50–53. IEEE Computer Society (2009)

42. You, Y.: Parallel ant system for traveling salesman problem on GPUs. In: Genetic and Evolutionary Computation, GECCO 2009, pp. 1–2 (2009)

43. Yu, Q., Chen, C., Pan, Z.: Parallel genetic algorithms on programmable graphics hardware. In: Advances in Natural Computation. LNCS, vol. 3612, pp. 1051–1059. Springer, Heidelberg (2005)

44. Zhu, W., Curry, J.: Parallel ant colony for nonlinear function optimization with graphics hardware acceleration. In: Proceedings of the 2009 IEEE International Conference on Systems, Man and Cybernetics, pp. 1803–1808. IEEE Press (2009)

Study of the Effect of Evaporation Temperature on the Qualitative and Quantitative Separation of Marine Mineral Salts: Prediction and Application on the Seawaters of the Atlantic Sea, Morocco

F. Z. Karmil[1,2(✉)], S. Mountadar[1,2], A. Rich[2], M. Siniti[2], and M. Mountadar[1]

[1] Laboratory of Water and Environment, Faculty of Sciences, Chouaib Doukkali University, 20, 24000 El Jadida, Morocco
fatimazahrakarmil@gmail.com
[2] Team of Thermodynamics Surface and Catalysis, Faculty of Sciences, Chouaib Doukkali University, 20, 24000 El Jadida, Morocco

Abstract. This work aims to determine the temperature effect on the separation of sea- salts mineral on the Moroccan Atlantic coast. The main objective of this study is the determination of the optimal conditions of the evaporation process through the prediction of the liquid-solid equilibrium of seawaters at each stage of the evaporation process. Hence, the physicochemical and thermodynamic properties of evaporation were exploited through the calculation codes Frezchem and Phreeqcl3 to quantify the effect of the ionic composition of raw and concentrated seawater on the evaluation of salinity and on the qualitative and quantitative separation of salts. Moreover, the results obtained by the used calculation codes were validated by experiments. The results obtained from simulation modeling for different temperatures (25 °C, 50 °C, 75 °C and 100 °C) and different compositions were in good agreement with the experimental results carried out under the determined optimal conditions. In addition, it can be concluded from the results that the saturation and quantitative separation thresholds of each type of salt for different concentrations of marine waters are different and this is mainly due to the variation in the solubility of each salt as a function of the variation in temperature. Concerning the temperature effect, it was observed that the quantity and quality of precipitated marine salts by evaporation of the raw and concentrated waters of the Atlantic Ocean varied with the studied temperatures.

Keywords: Seawater · Evaporation · Sea-salts

1 Introduction

Seawater is a complex multi-component system where almost all chemical elements are present in different concentrations (macro, micro and trace) and forms (simple and complex ions, dissolved gases, etc...). However, seawater can be used as a source of extraction of useful substances that can be resulting from different seawater

© Springer Nature Switzerland AG 2019
M. Ezziyyani (Ed.): AI2SD 2018, AISC 913, pp. 147–161, 2019.
https://doi.org/10.1007/978-3-030-11881-5_13

desalination processes such as membrane desalination, reverse osmosis, evaporation processes, and evaporation–cooling processes [1].

Evaporation is one of the most effective processes for the separation of dissolved salts present in seawater. It allows the study of marine evaporative systems. It depends on the operating conditions such as temperature (evaporation and air), air movement, salinity, etc. It requires well-defined separation operations to achieve extraction of mineral salts at different temperatures, and this makes this process relatively complex.

Several studies were done in order to understand the experimental and theoretical approach of evaporation process and deduce the effect of temperature on the separation of sea-salts. For example, [38] has studied the French coast of the Mediterranean with a salinity of 38.45‰ at T = 40 °C. He described the process followed and the order of precipitated salts in a quantitative manner according to the term "Usiglio sequence". This experience was confirmed by other studies: [2–4] which show that the first stage of evaporation is characterized by gypsum precipitation followed by halite and ended by K-Mg salts for high brine concentrations [5–9], all of whose results were made at different evaporation and salinity temperature conditions and were confirmed mainly by a set of theoretical calculations. Moreover, crystallization depends not only on the evaporation temperature but also on the kinetic modalities, the nature of the equilibrium established between the precipitated phases and the ionic composition [10–14].

The thermodynamic study quantifies the effect of the ionic composition and salinity of seawater on the precipitated mineral phases as a function of the volume of water evaporated at different temperatures. The theoretical sequence of fractional crystallization was developed by Jacobus Henricus vant'Hoff (1900) from the determination of solubility as well as salt saturation points. Pitzer calculation of liquid-solid equilibrium was based on the ion interaction model that was well developed by Marion et al.

In the present work, the physicochemical and thermodynamic properties of liquid-solid equilibrium of solutions concentrated at different temperatures on the Rabat coast and El Jadida Bay were studied. Knowledge of these properties is essential for studying the seawater evaporation process and interpreting the precipitation-crystallization phenomena involved. This includes interpreting and better defining the separation volumes and salt saturation thresholds validated by the Pitzer model at different temperatures.

2   Materials and Methods

2.1   Sample Pre-treatment and Concentration of Seawater by Evaporation

The carbonates were removed by adding H_2SO_4 up to pH = 7 and the seawater is concentrated by reduction of volume by evaporation process to different percentages (10%, 20%, 30%, 40%, 50%, 60%, 70%, 80%, 90% and 100%).

2.2   Chemical Analysis

Chemical analysis was applied to determine the composition of solution at each stage of the evaporation process of seawaters. Potentiometric titration was used to determine

Ca^{2+}, Mg^{2+} and Cl^- ions. The Ca^{2+} and Mg^{2+} ions concentrations were determined by EDTA complexmetric titration. The Cl^- ion concentration was determined argentometrically by $AgNO_3$. Thermogravimetric method was used for the determination of SO_4^{2-} with $BaCl_2$. The K^+ ion with sodium tetraphenylborate and of Na^+ ion was analysed by flame spectrometer respectively. The accuracy of these analyses was about 0.1–0.2% [1, 15].

2.3   Prediction of Liquid-Solid Equilibrium in Seawater

The liquid - solid equilibrium of seawater were evaluated by thermodynamic ion-interaction Pitzer model by using the Frezchem calculation code, at different temperatures (25 °C, 50 °C, 75 °C and 100 °C) considering only the six major elements of seawater (Na^+, Ca^{2+}, Mg^{2+}, K^+, Cl^- and SO_4^{2-}).

A thermodynamic ion-interaction was used the Frezchem (FREEZING CHEMISTRY) calculation code applied by Marion et al. [16–28]. It can also estimate the liquid-solid equilibrium of seawater during evaporation process [29].

Pitzer Approach

The Pitzer model [30–34] was based on Debye-Huckel extended law in which a set of terms have been added to determine the ionic strength (γ) of binary and ternary systems as a function of temperature at a pressure of 1.01 bar for cations (M), anions (X), and aqueous neutral species (N) [16–28]:

$$\ln(\gamma M) = Z_M^2 F + \sum m_a (2B_{Ma} + ZC_{Ma}) + \sum m_c \left(2\Phi_{Mc} + \sum m_a \Psi_{Mca} \right) + \\ \sum\sum m_a m_{a'} \Psi_{Maa'} + Z_M \times \sum\sum m_c m_a C_{ca} + 2\sum m_n \lambda_{nM} + \\ \sum\sum m_n m_a \zeta_{nMa} \tag{1}$$

$$\ln(\gamma X) = Z_X^2 F + \sum m_c (2B_{cX} + ZC_{cX}) + \sum m_a \left(2\Phi_{Xa} + \sum m_c \Psi_{cXa} \right) + \\ \sum\sum m_c m_{c'} \Psi_{cc'X} + |Z_X| \times \sum\sum m_c m_a C_{ca} + 2\sum m_n \lambda_{nX} + \sum\sum m_n m_c \zeta_{ncX} \tag{2}$$

$$\ln(\gamma N) = \sum m_c (2B\lambda_{Nc}) + \sum m_a (2\lambda_{Na}) \sum\sum m_c m_a \zeta_{Nca} \tag{3}$$

Where B, C, Φ, Ψ, λ and ζ are the interaction parameters of the Pitzer equation, mi is the molar concentration, with F and Z are the functions of the equation. In these equations, the Pitzer interaction parameters and the F function are temperature dependent. The indices c, a, and n refer to cations, anions, and neutral species, respectively. For c' and a' are referred to cations and anions that differ from c and a. Water activity (aw) at P = 1.01 bar is given by:

$$\alpha_w = \exp\left(-\frac{\emptyset \sum m_i}{55,50844} \right) \tag{4}$$

Where ø is the osmotic coefficient of which:

$$(\phi - 1) = \frac{2}{\sum_i m_i} \left\{ -\frac{A_\phi I^{3/2}}{1 + bI^{1/2}} + \sum \sum m_c m_a \left(B_{ca}^\phi + ZC_{ca} \right) + \sum \sum m_c m_{c'} \left(\Phi_{cc'}^\phi + \right. \right.$$
$$\left. \sum m_a \Psi_{cc'a} \right) + \sum \sum m_a m_{a'} \left(\Phi_{aa'}^\phi + \sum m_c \Psi_{caa'} \right) +$$
$$\left. \sum \sum m_n m_c \lambda_{nc} + \sum \sum m_n m_a \lambda_{nc} + \sum \sum \sum m_n m_c m_a \zeta n, c, a \right\} \quad (5)$$

Similarly, the temperature dependent as a function of changes in the parameters of the Pitzer equation and the solubility range of each product studied:

$$P = a_1 + a_2 T + a_3 T^2 + a_4 T^3 + \frac{a_5}{T} + a_6 \ln(T) \quad (6)$$

Where P is the Pitzer parameter, Ks is the product of solubility, T is the absolute temperature and a is the coefficient of interaction [16–28].

The equilibrium constant of the dissolution reaction of an ionic compound in water is called the solubility product Ks. When equilibrium is reached, the solution is said to be saturated:

$$M_m X_{n \, solide} \Leftrightarrow m M_{aq}^{z_{M+}} + n X_{aq}^{z_{X-}} \quad (7)$$

$$K_S = \frac{a_M^m a_X^n}{a_{M_m X_n(S)}} \qquad \text{With} \quad a_{M_m X_n(S)} = 1 \quad (8)$$

We can then write:

$$K_S = a_M^m a_X^n \quad (9)$$

Thermodynamic Simulation of the Precipitation of Sea Salts

Thermodynamic simulation of ionic interactions was performed using PHREEQC computer program, version 3, of major elements Na^+, K^+, Ca^{2+}, Mg^{2+}, Cl^-, and SO_4^{2-} to different temperatures. Mineral precipitation was simulated by using the Pitzer model. Due to the system complexity, the saturation index (SI) of precipitated solid phases, under experimental conditions, was calculated according to the following relationship:

$$SI = \log_{10}(IAP/Ksp) \quad (10)$$

Where IAP is the ionic activity product and K is the solubility product.

The calculations were carried out on the seawater compositions of Rabat and El Jadida at different temperatures 25 °C, 50 °C, 75 °C and 100 °C. The saturation index is used to determine the water is saturated, undersaturated, or supersaturated with respect to the state of the mineral [1].

3 Results and Discussions

3.1 Physicochemical Analysis After Evaporation

The analyses of the different samples prepared from the seawater of El Jadida bay and Rabat coast are grouped in Tables 1and 2.

The sum of the ions of each composition allows us to determine the relative salinity. Thus the salinity of the seawaters of El Jadida Bay is estimated at 36.34 g/kg and on the Rabat coast at 34.07 g/kg for the raw sample. Moreover, salinity increases with the evaporation percentage of seawater. For each physicochemical parameter, it is noted that it has an elevation proportional to evaporation rates [35]. Indeed, the distribution of the different ions is practically the same (Cl^- (50%), Na^+ (29%), SO_4^{2-} (12%), Mg^{2+} (4%), K^+ (2%), Ca^{2+} (3%)) [36].

Table 1. Composition of raw seawater from El Jadida bay and Rabat coast

Raw seawater		
	El Jadida	Rabat
X(mS/cm)	60.8	51.6
pH	8.08	8.12
$[Ca^{2+}](g/l)$	1.203	0.191
$[Mg^{2+}](g/l)$	1.531	0.67
$[Cl^-](g/l)$	18.366	20.17
$[SO_4^{2-}](g/l)$	4.51	1.412
$[K^+](g/l)$	0.011	0.013
$[Na^+](g/l)$	10.72	11.614
S(g/l)	36.34	34.07

Table 2. Composition of concentrated seawater from El Jadida bay

	The % volume of water remaining after evaporation of El Jadida seawater								
	10%	20%	30%	40%	50%	60%	70%	80%	90%
X(mS/cm)	180.2	177.3	148.2	124	117.8	98.5	84.6	76.5	71
pH	8.02	7.99	7.98	7.95	7.93	7.91	7.89	7.86	7.8
$[Ca^{2+}](g/l)$	6.13	5.654	5.49	5.13	4.5	3.85	3.21	2.53	1.93
$[Mg^{2+}](g/l)$	2.38	2.284	2.23	2.06	1.99	1.97	1.88	1.8	1.75
$[Cl^-](g/l)$	75.26	63.19	51.83	38	34.8	32.66	22.01	19.7	18.1
$[SO_4^{2-}](g/l)$	12.87	12.4	11.53	9.22	8	7.2	7.12	5.35	4.76
$[K^+](g/l)$	1.19	1.04	0.85	0.73	0.61	0.55	0.43	0.33	0.24
$[Na^+](g/l)$	56.50	48.89	41.39	32.19	32.22	29.02	22.51	12.63	12.17
S (g/l)	154.34	133.45	113.3	87.33	82.08	75.24	57.15	42.34	38.95

3.2   Effect of Evaporating Temperature on the Qualitative and Quantitative Separation of the Different Solid Phases

The solubility of a salt in a solvent varies with temperature, according to the general law of equilibrium. The rise in temperature will cause an increase in solubility. This is what is most often observed: the solubility of many solids phases increases with elevation of temperature [36, 37].

The First Stage of Separation

First, gypsum or calcium sulphate (CaSO4.2H2O) is the first mineralogical phase precipitated. This is explained by its low solubility and rapid crystallization.

For T = 25 °C, calcium sulphate precipitation is carried out at a salinity interval between 155.66 and 306.50 g/Kg and a remaining volume of 14.22%. The maximum mass is 0.98 g for each liter of seawater. For T = 50 °C, the precipitation of calcium sulphate is deposited during a salt concentration varies between 162.33 and 333.54 g/kg and a remaining volume is 13.25%. The maximum mass is 0.92 g for each liter of seawater. For T = 75 °C, calcium sulphate precipitation is carried out at a salt concentration ranging between 163.71 and 352.18 g/Kg and a remaining volume of 12.34%. The maximum mass is 0.87 g for each liter of seawater. For T = 100 °C, the precipitation of calcium sulphate is deposited at a salinity ranging from 180 to 365.44 g/kg and the remaining volume is 12.07%. The maximum mass is 0.82 g for each liter of sea water (Fig. 1).

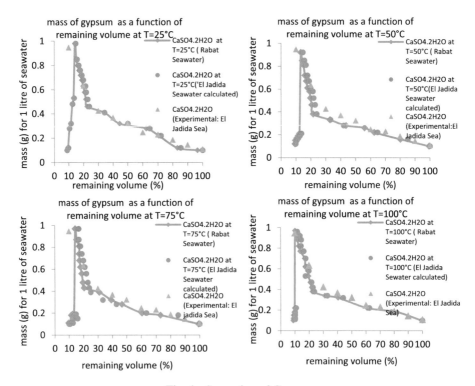

Fig. 1. Separation of Gypsum

In some studies, gypsum crystallization begins from 150 g/Kg and continues until the start of halite crystallization at a varying salinity of 290 to 320 g/kg at a seawater density of 1.20 to 1.26 and a reduction volume of 20% at T = 40°C at the Mediterranean coast [39] and 19% at Macleod, west Australia [3].

The Second Stage of Separation

Then, sodium chloride (NaCl) is the second mineralogical phase precipitated after gypsum (Fig. 2).

For T = 25 °C, the precipitation of sodium chloride is carried out at a salinity range between 306.50 and 401 g/Kg and a remaining volume of 9.75% The maximum mass is 22.357 g for each liter of sea water.

For T = 50 °C, the precipitation of sodium chloride is deposited during a salt concentration varies between 310 and 400 g/l and a remaining volume of 9.25%. The maximum mass is 22.26 g for one liter of seawater.

For T = 75 °C, the precipitation of sodium chloride is carried out at a salt concentration ranging between 315 and 426 g/l and a remaining volume of 9.05%. The maximum mass is 21.82 g for each liter of sea water. For T = 100 °C, sodium chloride precipitation is deposited at a salinity between 318 and 460 g/kg and a remaining volume of 8.85%. The maximum mass is 21.6 g for each liter of sea water.

As highlighted by Usiglio and Logan, halite begins to crystallize when seawater is evaporated from 10% to 8% of the original volume. Halite unlike all other minerals it requires a relatively low degree of super-saturation to begin to precipitate [3, 14, 39].

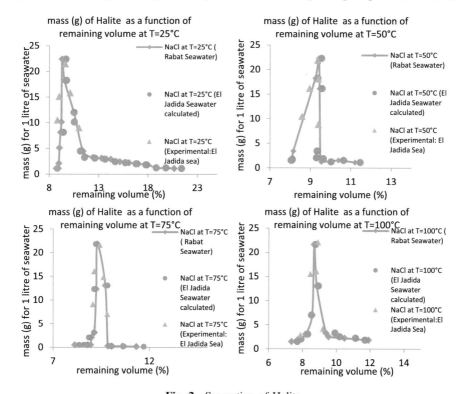

Fig. 2. Separation of Halite

The Third Stage of Separation

The reduction of seawater volume at different temperatures causes the crystallization of bloedite ($Na_2SO_4.MgSO_4.4H_2O$).

For T = 25 °C, the precipitation of sodium and magnesium sulphate is carried out at a salinity interval between 306.54 and 387.45 g/Kg at V = 7.5% and m_{max} = 0.68 g. For T = 50 °C, the precipitation of sodium and magnesium sulphate is deposited in a salt concentration between 333.54 and 389.24 g/l at V = 6.90% and m_{max} = 0.6 g. For T = 75 °C, the precipitation of sodium and magnesium sulphate is carried out at a salt concentration ranging between 345.33 and 396.73 g/l at V = 6.8% and m_{max} = 0.58 g. For T = 100 °C, the precipitation of sodium and magnesium sulphate is deposited at a salinity greater than 360 g/kg at V = 6.5% and m_{max} = 0.50 g (Fig. 3). Some studies show that after gypsum precipitation, halite began to crystallize followed by bloedite [2, 4–6, 8, 9, 14].

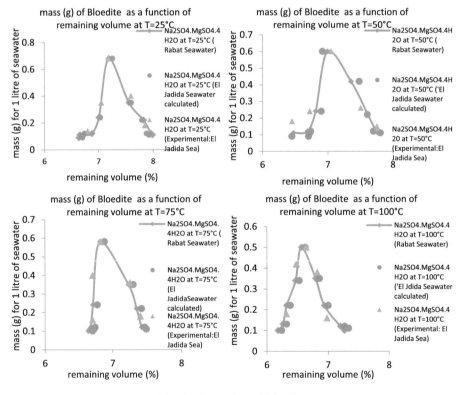

Fig. 3. Separation of Bloedite

In this work the precipitation of the bloedite in parallel with the halite is explained by the effect of the ionic composition of the seawater treated in the different studies. In the Adriatic Sea, the first precipitate is gypsum at V = 13% and extends to the halite super saturation range [6].

Thus, the crystallization of halite continues to a very high salinity, passing through the point where the magnesium sulphate crystallizes. As a result, the 3 minerals crystallize together at a salinity of 375 g/kg and a brine density of 1.32 [14].

The Last Stage of Separation

Finally, K-Mg salts are the last precipitates, For T = 25 °C, precipitation of K-Mg type salts is carried out at a salinity interval between 380 and 450 g/Kg at V = 2.90% and m_{max} = 0.26 g.

For T = 50 °C, the precipitation of K-Mg is deposited during a salt concentration varies between 380 and 455 g/l at V = 2.46% and m_{max} = 0.2 g. For T = 75 °C, precipitation of K-Mg is carried out at a salt concentration ranging between 350 and 450 g/l at V = 1.98% and m_{max} = 0.17 g. For T = 100 °C, the precipitation of K-Mg is deposited at a salinity of 350 to 500 g/kg and a remaining volume of 1.48%. The maximum mass is 0.15 g for each liter of sea water (Fig. 4).

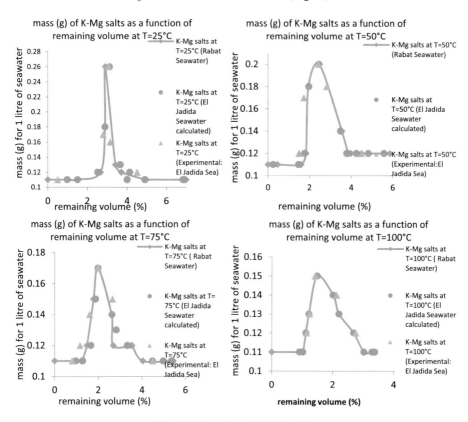

Fig. 4. Separation of K-Mg salts

The different precipitated K-Mg salts have been extensively studied by different authors [39, 40]. They confirm that precipitation of K-Mg salts is carried out at very advanced stages of seawater concentration by evaporation. It extends to the last drop in the solution. As for [40], the K-Mg sequence produced during the typical temperature fluctuations of the Crimean coast on the Black Sea described can be divided into 4 phases. It begins with the precipitation of Mg or Na–Mg sulphate without K (without sylvinite and carnallite). Then precipitation of K-Mg salts (in particular sylvite without carnallite). After, we can see the precipitation of carnallite and finally bischofite in the terminal stage [14]. It should be noted that in the present work only the effect of temperature on the main precipitated phases is considered and do not have the determination of the detailed evaporative sequence at the Atlantic coast levels.

All empirical observations show a set of inconsistencies during the later stages of evaporation, mainly related to the influence of different temperatures, appear in the later stages of precipitation of K-Mg type salts where brines are highly concentrated [2, 4–6, 8, 9, 14]. The present work is characterized by gypsum precipitation followed with Halite (NaCl) then bloedite ($Na_2SO_2.MgSO_4.2H_2O$) and at the end K-Mg salts.

This is explained not only by the effect of the different temperatures on the ionic composition, but also by the kinetic nucleation and crystallization factors used to define the stability during the equilibrium between the precipitated mineral phases [35–41]. Therefore, several evaporative crystallization paths are applicable. As a result, several theoretical evaporite sequences will be produced and all are possible depending on the study area [14].

3.3   Comparison Between the Results Obtained and the Prediction of the Calculation Mode at Different Evaporation Temperatures

The Effect of Temperature on Precipitated Salt Concentrations During Evaporation

During evaporation, the correspondence between the volume reduction rate and the global salt concentrations at different temperatures does not follow a simple law.

Indeed, when a salt crystallizes (in variable quantity according to the type of salt), it modifies the concentration of ions which remain in solution.

For each reduction of seawater volume the ions concentrate in the remaining solution (Fig. 5).

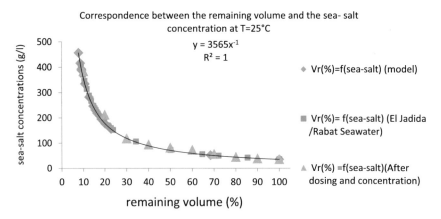

Fig. 5. Correspondence between volume (%) and sea-salt mineral

Moreover this concentration of ions evolves in a linear way until a level varying from 180 to 380 g/l where the biggest part of the salts crystallizes and this corresponds to a volume varying between 0% and 20%. As a result, we note on the one hand the validation of the calculation code and on the other hand the same types of salinity evolution as a function of the % volume reduction and this for the different temperatures [36].

The evaporation of seawater causes a decrease in the volume of water which is accompanied by an increase in the concentration of solutes (mineral salts in our case). If the concentration exceeds the solubility of the solutes then they precipitate. The precipitation of mineral salts decreases when the temperature of the solution increases according to the general law of liquid-solid equilibrium.

In this study, the correspondence between the volume reduction rate remaining as a function of salt concentrations shows the same rates with a slight variation (<1%) over the different temperatures: 25 °C; 50 °C; 75 °C and 100 °C.

Under the effect of temperature, the rate of loss between T = 25 °C and T = 100 °C is probably related to the activity of mineral salts precipitated in seawater (Fig. 6).

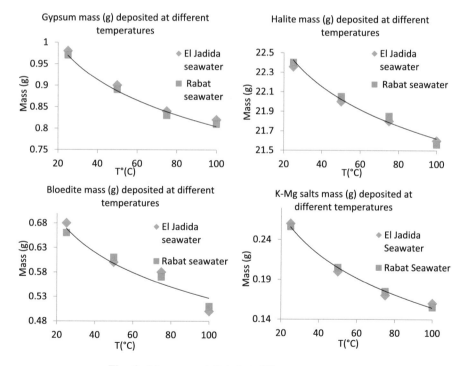

Fig. 6. Masses precipitated at different temperatures

For gypsum ($CaSO_4.2H_2O$), the degree of mass loss between T = 25 °C and T = 100 °C is 16.3%. For halite (NaCl), the degree of mass loss between T = 25 °C and T = 100 °C is 3.4%. For bloedite ($Na_2SO_4.MgSO_4.4H_2O$), the degree of mass loss between T = 25 °C and T = 100 °C is 26.47%. For K-Mg salts, the degree of mass loss between T = 25 °C and T = 100 °C is 42% so note the high sensitivity of K-Mg salts to high temperature. It can therefore be deduced that the best separation temperature for marine mineral salts corresponds to natural conditions (T = 25 °C and P = 1 atm). As for Braitsch (1971, 1978), the optimal temperature for a high mineral salt yield at different crystallization paths is T = 25 °C. In this work, the salts are separated by evaporation process, respecting the saturation intervals of each salt. Temperature influences yields and the higher the temperature the faster the crystallization rate increases. At T = 100 °C the total evaporation is obtained after a few hours while at T = 25 °C it is reached after a few weeks. Moreover, the Frezchem calculation code describes very well the different liquid-solid equilibrium of the different salts that crystallize as the overall volume of seawater is reduced. Therefore, the calculation code was performed to evaluate the impact of temperature on the crystallization of the different salts.

Saturation Index
Figure 7 shows that thermodynamic calculation of the saturation index of precipitated solid phases at T = 25 °C confirm the previous results. They revealed that the four main phases have a positive saturation index and could precipitate to an evaporation volume of 90% to 100%. With the exception of gypsum which begins to precipitated as soon as the volume of evaporation is 15% but in very small quantities. This is explained by their low solubility Ks = 3.14 * 10^{-5} at T = 25 °C.

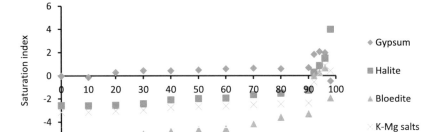

Fig. 7. Saturation indices for different sea-salts at T = 25°C

This allows calcium sulphate to reach saturation very quickly and crystallize very early compared to other salts. However, values above 0 for other salts do not automatically cause precipitation of a mineral phase, because precipitation kinetics are very slow at T = 25 °C.

These results show that these minerals cannot be precipitated from the initial seawater solution. This implies that the brine must be concentrated many times before saturation in relation to the soluble salts produced. It is in this context a set of isothermal evaporation experiments were carried out, in order to determine the nature and precipitation sequence of minerals dissolved in seawater [1].

4   Conclusion

In this work, the effect of temperature on the quality and quantity of precipitated mineral sea-salts were studied and demonstrated experimentally and predicted using the Frezchem and PhreeqcI3 calculation codes. However, the determination of the liquid-solid equilibrium shows that the saturation thresholds for each type of sea-salt are different and the limit concentration of each salt depends on its solubility as a function of temperature variation. The results obtained from simulation modeling for different temperatures (25 °C, 50 °C, 75 °C and 100 °C) and for the different compositions were in good agreement with the experimental results carried out under the determined optimal conditions.

Moreover, it was observed that the determined quantities of different sea-salts obtained namely $CaSO_4.2H_2O$, $NaCl$, $Na_2SO_4.MgSO_4.4H_2O$ and K-Mg salts varied at different temperatures. Mineral salt yields during crystallization are likely related to evaporation temperature and its variations. As the temperature increases, the rate of mass loss increases between T = 25 °C and T = 100 °C. For gypsum ($CaSO_4.2H_2O$) the loss degree is 16.3%. For halite ($NaCl$) loss degree is 3.4%. The bloedite ($Na_2SO_4.MgSO_4.4H_2O$) the loss degree is 20%, and for K-Mg salts, the degree of loss is 42%. Therefore, the optimal temperature for a high mineral salt yield at different crystallization paths is T = 25 °C.

The study of the different sea-salts precipitated during evaporation processes will allow the valorisation, in terms of quantity and quality, of the salts resulting by separation and purification. Also the processes will be realized for the others brines resulting by desalination techniques. The following of this study will allowed at different pre-treatment of the seawater before the evaporation processes for improvement of quality and quantity of different salts precipitated.

References

1. Kovacheva, A., Rabadjieva, D., Tepavitcharova, S.: Simulation of stable and metastable sea-type carbonate systems for optimization of MgCO3.3H2O precipitation from waste sea brines. Desalination **348**, 66–73 (2014)
2. Bassegio, G.: The composition of sea water and its concentrates. In: Coogan, A.H. (ed.) 4th International Symposium on Salt, Houston, Texas, 8–12 April, vol. 2, pp. 351–358 (1974)
3. Logan, B.W.: The MacLeod Evaporite Basin, Western Australia: Holocene Environments, Sediments and Geological Evolution, 44, pp. 1–140. AAPG Memoir (1987)
4. McCaffrey, M.A., Lazar, B., Holland, H.D.: The evaporation path of seawater and co precipitation of Br⁻ and Kþ with halite. J. Sediment. Petrol. **57**, 928–937 (1987)
5. Garrett, D.E.: The chemistry and origin of potash deposits. In: Rau, J.L., Dellwig, L.F. (eds.) 3rd Symposium on Salt, Cleveland, 21–24 April, vol. 1, pp. 211–222. Northern Ohio Geological Society, Cleveland, OH (1969)
6. Herrmann, A.G., Knake, D., Schneider, J., Peters, H.: Geochemistry of modern seawater and brines from salt pans: main components and bromine distribution. Contrib. Miner. Petrol. **40**, 1–24 (1973)

7. Busson, G., Cornée, A., Dulau, N., et al.: Nature et genèse des faciès confinés: Données hydrochimiques, biologiques, isotopiques, sédimentologique et diagénétiques sur les marais salants de Salin-de-Giraud (Sud de la France). Nature and genesis of confined facies: Hydrochemical, biological, isotopic, sedimentological and diagenetic data on the salt marshes of salin in Giraud (South of France). Mediterr. Geol. **9**, 303–592 (1982)
8. Orti, F., Busson, G.: Introduction to the sedimentology of the coastal salinas of Santa Pola (Alicante, Spain). Rev. Inv. Geol. **38–39**, 9–229 (1984)
9. Geisler-Cussey, D.: Approche sédimentologique et géochimique des mécanismes générateurs de formations évaporitiques actuelles et fossiles. Marais salants de Camargue et du Levant espagnol, Messinien méditerranéen et Trias lorrain. Sedimentological and geochemical approach of the evaporite and fossil formations: Camargue and Spanish levant salt marshes, Mediterranean Messinian and Lorraine triassic. Sciences de la Terre, Nancy: Fondation scientifique de la géologie et de ses applications. Mémoires, vol. 48, pp. 1–268 (1986)
10. Valyashko, M.G.: Geochemical Rules of the Potassium Salt Deposits Formation, pp. 1–398. Izdatelstvo Moskovskovo Universiteta, Moscow (1962). (in Russian)
11. Krauskopf, K.B.: Introduction to Geochemistry, pp. 1–721. McGraw-Hill/Kōgakusha, New York/Tokyo (1967)
12. Garrett, D.E.: Potash Deposits, Processing, Properties and Uses. Chapman and Hall, London (1996)
13. Jadhav, M.H.: Recovery of crystalline magnesium chloride-hexahydrate by solar evaporation of sea bitterns. In: Schreiber, B.C., Harner, H.R. (eds.) 6th International Symposium on Salt, Toronto, Ontario, Canada, 24–28 May, vol. 2, pp. 417–419 (1983)
14. Bąbel, M., Schreiber, B.C.: Geochemistry of Evaporites and Evolution of Seawater (2014)
15. AFNOR, Recueil des normes Françaises: Qualité de l'eau, 3e édn/Collection of Frensh Standards: Water quality, 3rd edn (1999)
16. Marion, G.M., Farren, R.E.: Mineral solubilities in the Na–K–Mg–Ca–Cl–SO 4–H2O system: a re-evaluation of the sulfate chemistry in the Spencer-Møller-Weare model. Geochim. Cosmochim. Acta **63**, 1305–1318 (1999)
17. Marion, G.M., Kargel, J.S.: Cold Aqueous Planetary Geochemistry with FREZCHEM: From Modeling to the Search for Life at the Limits. Springer, Berlin (2008)
18. Marion, G.M., Catling, D.C., Kargel, J.S.: Br/Cl partitioning in chloride minerals in the Burns formation on Mars. Icarus **200**, 436–445 (2009)
19. Marion, G.M., Catling, D.C., Crowley, J.K., Kargel, J.S.: Modeling hot spring chemistries with applications to martian silica formation. Icarus **212**, 629–642 (2011)
20. Marion, G.M., Catling, D.C., Zahnle, K.J., Claire, M.W.: Modeling aqueous perchlorate chemistries with applications to Mars. Icarus **207**, 675–685 (2010)
21. Marion, G.M., et al.: Modeling aluminum-silicon chemistries and application to Australian acidic playa lakes as analogues for Mars. Geochim. Cosmochim. Acta **73**, 3493–3511 (2009)
22. Marion, G.M., Kargel, J.S., Catling, D.C.: Modeling ferrous-ferric iron chemistry with application to Martian surface geochemistry. Geochim. Cosmochim. Acta **72**, 242–266 (2008)
23. Marion, G.M., Kargel, J.S., Catling, D.C., Lunine, J.I.: Modeling ammonia-ammonium aqueous chemistries in the Solar System's icy bodies. Icarus **220**, 932–946 (2012)
24. Marion, G.M., Kargel, J.S., Catling, D.C., Lunine, J.I.: Modeling nitrogen–gas,-liquid,-solid chemistries at low temperatures (173–298 K) with applications to Titan. Icarus **236**, 1–8 (2014)
25. Marion, G.M., Kargel, J.S., Catling, D.C., Lunine, J.I.: Sulfite–sulfide–sulfate—carbonates equilibria with applications to Mars. Icarus **225**, 342–351 (2013)

26. Marion, G.M., Kargel, J.S., Tan, S.P.: Modeling nitrogen and methane with ethane and propane gas hydrates at low temperatures (173–290 K) with applications to Titan. Icarus **257**, 355–361 (2015)
27. Marion, G.M., Mironenko, M.V., Roberts, M.W.: FREZCHEM: a geochemical model for cold aqueous solutions. Comput. Geosci. **36**, 10–15 (2010)
28. Marion, G.M., Catling, D.C., Kargel, J.S., Crowley, J.K.: Modeling calcium sulfate chemistries with applications to Mars. Icarus **278**, 31–37 (2016)
29. Rich, A.: Dessalement de l'eau de mer par congélation sur paroi froide: aspect thermodynamique et influence des conditions opératoires, Seawater desalination by freezing on cold walls: thermodynamic aspect and influence of operating conditions. University of Rabat (2011)
30. Pitzer, K.S.: Thermodynamics of electrolytes. I. Theoretical basis and general equations. J. Phys. Chem. **77**(2), 268 (1973)
31. Pitzer, K.S., Mayorga, G.: Thermodynamics of electrolytes. III. Activity and osmotic coefficients for 2–2 electrolytes. J. Solution Chem. **3**(7), 539 (1974)
32. Pitzer, K.S., Mayorga, G.: Thermodynamics of electrolytes. II. Activity and osmotic coefficients for strong electrolytes with one or both ions univalent. J. Phys. Chem. **77**(19), 2300 (1973)
33. Pitzer, K.S.: Ion interaction approach: theory and data correlation. In: Pitzer, K.S. (ed.) Activity Coefficients in Electrolyte Solutions, 2nd edn, pp. 75–153. CRC Press, Boca Raton (1991)
34. Pitzer, K.S.: Thermodynamics, 3rd edn. McGraw-Hill, New York (1995)
35. Droubi, A., Fritz, B., Tardy, Y.: Equilibres entre minéraux et solutions: programmes de calcul appliqués à la prédiction de la salure des sols et des doses optimales d'irrigation. Equilibrium between minerals and solutions: calculation program applied to the prediction of soil salinity and optimal irrigation doses. 19 (1979)
36. Paul, J., Rivron, J.P.: La cristallisation fractionnée des sels marins: approche théorique de l'évaporation. Fractionated crystallization of sea-salts, December 2012
37. Weast, R.C., et al.: Handbook of Chemistry and Physics, 51st edn. CRC, Cleveland (1971)
38. Usiglio, J.: Analyse de l'eau de la Méditerranée sur les côtes de France. Mediterranean water analyses on the Coast of France. Ann. Chim. Phys. **3**(27), 92–107 (1849)
39. Usiglio, J.: Etudes sur la composition de l'eau de la Méditerranée et sur l'exploitation des sels qu'elle contient. Mediterranean and on the exploitation of the salts it contains. Ann. Chim. Phys. **3**(27), 172–191 (1849)
40. Braitsch, O.: Salt Deposits: Their Origin and Composition, pp. 1–297. Springer, Berlin (1971). (Translated from German edition, Springer, Berlin 1962, updated by A. G. Herrmann)
41. Bea, S.A., Carrera, J., Ayora, C., Batle, F.: Modeling of concentrated aqueous solutions: efficient implementation of Pitzer equation in geochemical and reactive transport models. Comp. Geosci. **36**, 526–538 (2010)

Performance of *Aspergillus niger* and *Kluyveromyces marxianus* for Optimized Bioethanol Production from Dairy Waste

Dounia Azzouni[1]([⊠]), Amal Lahkimi[2], Bouchra Louaste[3],
Mustapha Taleb[2], Mehdi Chaouch[1], and Noureddine Eloutassi[1]

[1] Process Engineering and Environment Laboratory,
Faculty of Sciences Dhar El Mahraz,
Sidi Mohamed Ben Abdellah University, Fez, Morocco
azzouni.dounia@gmail.com
[2] Engineering, Electrochemistry, Modelling and Environmental Laboratory,
Faculty of Sciences Dhar El Mahraz,
Sidi Mohamed Ben Abdellah University, Fez, Morocco
[3] Biotechnology Laboratory, Faculty of Sciences Dhar El Mahraz,
Sidi Mohamed Ben Abdellah University, Fez, Morocco

Abstract. Generally, biotechnology is used for a cleaner production and an increased energy effectiveness with less greenhouse gases emissions responsibles of serious environmental problems: air and water pollution.

Parallel to the increase of generated organic waste, during last decades, the recourse to renewable sources of energy receives considerable interest in the whole world. The energy of biomass is one of the promising sources, which contributes to organic waste valorization: wastewater, agricultural residues and industrial waste that can be used for bio-fuel production.

In this study we are interested in dairy waste valorisation as those effluents are characterized by high organic matter content, considerable concentrations of oils and greases, high levels of BOD_5, COD, nitrogen and phosphorus, significant variations of the pH and temperature and a high conductivity which largely exceeds those of domestic waste.

Thus, our study is devoted to the production of bioenergy from dairy waste by two yeast Strains (*Aspergillus niger and Kluyveromyces marxianus*) characterized by a great output of ethanol and a strong performance in the depollution of these industrial effluents.

Keywords: Bioenergy · Aspergillus niger · Kluyveromyces marxianus · Depollution · Bioethanol · Dairy waste

1 Introduction

Mainly, more than 80% of the world energy resources are fossil origins. The international recommendations urge all countries to reduce their greenhouse gas emissions. This obliges them to gradually move worms of the local sources of energy CO_2 neutrals [1–3].

© Springer Nature Switzerland AG 2019
M. Ezziyyani (Ed.): AI2SD 2018, AISC 913, pp. 162–175, 2019.
https://doi.org/10.1007/978-3-030-11881-5_14

The milk industries annually produce hundreds million kilograms of milk and its derivatives. This transformation involves the rejection of enormous quantities of residues, which present a considerable environmental problem [4–6]; the principal rejections are white water, the milk serum and the buttermilk. The strong BOD load of these rejections, higher than 50 g/l, constitutes a serious problem since its direct discharge into the treatment plants is not only expensive but overload the installations of treatment. However, the lactose contained in this medium represents 90% of the BOD load is a frightening factor of pollution [7–10].

Bioconversion of dairy waste into bioethanol would have environmental and economic advantages [11–14].

In this work, the lactoserum can be exploited for the production of bioethanol by fermentation after treatment of the organic matter. Then, the principal objectives of this work were summarized as follows: To produce monomeric sugars starting from the lactoserum by chemical and biological treatments. These sugars were used for the energy production, to study new strains in the process of bioconversion and to optimize the conditions of treatments.

2   Materials and Methods

2.1   Substrate

The substrate used is the lactoserum. In addition, we used simple sugars resulting from the physical-chemical and biological hydrolysis like substrates. The experiments were also carried out on pure lactose solutions, which were obtained by mixture of D-lactose (Ltd Sigma-Aldrich) with distilled water, and on lactose resulting from ultrafiltration. These substrates were stored at 4 °C until use.

2.2   Proportioning of the Parameters of the Effluent

In order to determine the impact of the rejections and to confirm the feasibility and the results of the treatments carried out, the composition of the lactoserum, the pH, the COD, the BOD_5 and ratio COD/BOD_5 were analyzed before and after valorization.

2.3   Proportioning of Lactose and Simple Sugars

Sugars in the various mediums are determined by the dinitrosalicylic method of the acid (DNS) described by Miller (1969). It is based on the formation of a chromatophore between the free carbonyl group of sugars and dinitrosalicylic acid (DNS). In a test tube of 20 ml containing 3 ml of Miller reagent, we added 2 ml of sample solution. After heating during 15 min with 100 °C and cooling at ambient temperature, we measured the absorbance at 640 nm. A calibration curve ranging from 0 to 1 g/l was prepared starting from a stock solution of 1 g/l of glucose. The method phenol - acid sulphuric was used to evaluate the spectrum of monomeric sugars resulting from the hydrolysis [15].

2.4   Micro-organisms

For the production of bioethanol, we used several mushroom and yeast Strains as mentioned in Table 1.

Table 1. Origin of Strains

Strain	Origin
Kluyveromyces marxianus *Candida inconspicua* *Candida xylopsoci*	Commercial ferment from the laboratory of food industry, Hassan II Institute of Agronomy & Veterinary Medicine – Rabat-Morocco
Aspergillus niger *Saccharomyces cerevisiae*	Laboratory of Biotechnology and Molecular Microbiology. Faculty of Science Dhar El Mahraz. Fez. Morocco

2.5   Culture Media

All fermentations carried out with the yeasts (*Kluyveromyces marxianus, Candida inconspicua, Candida xylopsoci, Saccharomyces cerevisiae*) were maintained in precultures on YEPD liquid medium (20 g/l of glucose, 20 g/l of peptone, 10 g/l of yeast extract, and sterile H_2O to 1000 ml) incubated at various temperatures. *Aspergillus niger* culture was carried out on the liquid Sisler medium which contains 2 g/l of KH_2PO_4, 1.5 g/l of K_2HPO_4, 1 g/l of $(NH_4)_2SO_4$, 0.5 g/l of $MgSO_47H_2O$ and 2 g/l of yeast extract and incubated at various temperatures.

2.6   Conditions of Fermentations

All the cultures of the strains used for ethanolic fermentation were carried out in a bioreactor of 2 L with a working volume of 1.2 L. The mediums used for the experiments contain various concentrations of substrates (raw lactoserum, lactose resulting from ultrafiltration, simple sugars resulting from the hydrolysis). In addition, we optimized the parameters influencing the processes of fermentation of the free cells.

- **Effect of the tolerance of the strains towards exogenic ethanol:** The tolerance towards exogenic ethanol of the studied Strains was measured by the percentage of survival of these microorganisms in the solid medium (YEPD for yeasts and Sisler for *Aspergillus niger*) supplemented by various ethanol concentrations.
- **Effect of the pH:** various values of pH (adjusted by the tartaric acid, 2N) were tested (30 °C, 15% of sugar) for *Aspergillus niger* and with 48 °C for yeasts.
- **Effect of the temperature** with the optimal value of found pH, we incubated the free cells at various temperatures in medium YPG (1% of the yeast extract, 2% of peptone, and 2.5% of glucose). The selected temperatures are 22 °C, 30 °C, 37 °C and 42 °C, 48 °C and 58 °C for the cells used.
- **Effect of the concentration in substrates:** In these experiments, we tried to determine the sugar concentration from which the production of ethanol is maximum. The proportioning of reducing sugars in the various mediums are determined by the dinitrosalicylic acid method (DNS) as described by Miller (1969).

- **Effect of the initial biomass:** The initial sugar concentration used for this study was 22 g/l, the pH was adjusted to 7 with the tartaric acid and the temperature was fixed at 30 °C for *Aspergillus niger* and 48 °C for yeasts, whereas various concentrations of inoculum of the free cells were tested on the production of ethanol.

2.7 Distillation and Proportioning of Ethanol

The ethanol was determined by comparing the refraction index of medium culture, with known ethanol concentrations, and a kit of ethanol proportioning Boehringer Mannheim.

2.8 Determination of the Dry Extract (Single Cell Proteins)

The single cell protein determination was ensured by the evaluation of the total dry extract measured every two hours. 10 ml of medium were centrifuged during 10 min with 5000 rpm at 4 °C. The sediment was washed twice-distilled water then placed at the drying oven at 105 °C during 24 h. A weighing with a balance of precision was carried out. Chemical and biochemical analysis were carried out on the supernatant of the culture.

3 Results and Discussion

3.1 Characteristics of the Lactoserum of Dairy Transformation

Several indicators are used to evaluate the pollution load of an effluent. The two following Tables (2 and 3) point out some required parameters of the effluent. In the lactoserum, the lactose represents 90% of the BOD load what constitutes at the same time, a substrate which may undergo beneficiation and a real threat for the environment.

Table 2. Physico-chemical characteristics of various dairy effluents (LVRS)

Type of the effluent	pH	Volume of product/l	COD (g/l)	COD/BOD$_5$
White water	5.5	3 at 4	2,5	1.3
Lactoserum	4.3	0,75	6	1.5
Total effluent	4.4	4 at 5	11	4.7

Consequently, the effluents of various dairy rejections must undergo a treatment to reduce their organic load. The rejections of LVRS are primarily made up of water, lactose, proteins, minerals and the fats (Table 3).

3.2 Lactose Hydrolysis

The lactose is a diholoside (or disaccharide), composed of a β-D-galactose molecule (Gal) and of a molecule of α/β-D-glucose (Glc) connected between them by an osidic connection β (1-4). The official name of lactose is the β-D-galactopyrannosyl (1-4) D

Table 3. Composition of lactoserum of the rejection of the processing industry of the LVRS

Components	Lactoserum
Lactose (g/100 gMS)	64
Proteins (g/kgMS)	82.5
Mineral-ashes (g/kgMS)	41
Organic acids (g/l)	11
Fat content (g/l)	2.1
Lactose (g/100 gMS)	64
Proteins (g/kgMS)	82.5

glucopyrannose. It can be symbolized by Gal β (1-4) Glc; the lactose contained in the lactoserum can be a good energy substrate according to several library searches. The reformation of lactose is of great interest for food industry. The lactose must initially be segmented in the presence of a water molecule into D-glucose and D-galactose (Fig. 1).

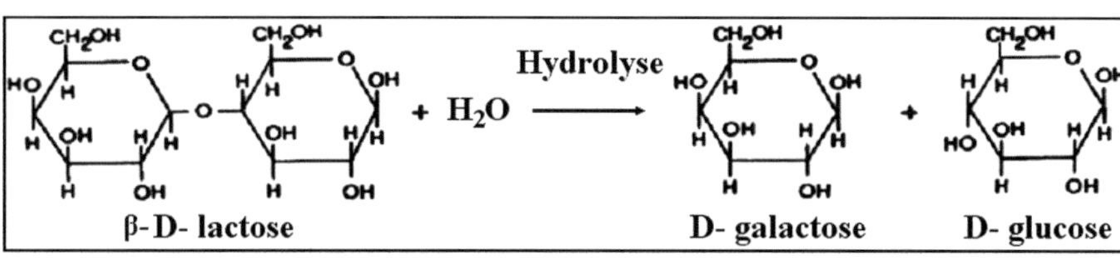

Fig. 1. Lactose hydrolysis in the presence of a water molecule

Several methods of lactose hydrolysis were carried out. The substrate was treated by the sulphuric acid, the hydrochloric acid and by the carbon dioxide [16, 17]. The concentration of the acid, the temperature and the processing time were optimized using response surface methodology [18, 19].

3.3   Valorization of the Lactoserum by the Production of Bioethanol by Free Cells

The general process of the valorization of the lactoserum for the production of bioethanol consists of three phases (Fig. 2). The first phase consists in filtering the lactoserum and white water by ultrafiltration to separate lactose from the other components. The second phase is the lactose hydrolysis to release fermentable simple sugars. The third phase is fermentation of these sugars for the production of bioethanol.

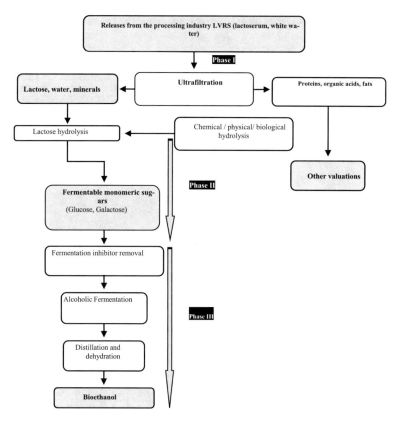

Fig. 2. Processes of production of bioethanol starting from the rejections of the processing industry of milk

3.4   Optimization of the Conditions of Alcoholic Fermentation

- Effect of the tolerance of the Strains of fermentation with exogenic ethanol

The tolerance with exogenic ethanol was measured by the percentage of survival of the Strains of alcoholic fermentation in their mediums of growth supplemented by various ethanol concentrations. The results were reported in Table 4.

It arises from the results of the analyses that the inhibition of the growth of all yeasts and the mushroom used in this work were weak for exogenic ethanol rates lower than 5% of which the percentage of growth and survival is higher than 50% and it is very high exceeding 90% at an ethanol rate of 4.5%. Beyond 5%, we noticed a fast fall of the growth rate for all the Strains, the size of the colonies decreases as the ethanol concentration increases, but with concentrations of 7% out of ethanol, *Saccharomyces cerevicea* and *Aspergillus niger* remain viable but their division is blocked. The examination of limp of Petri for ethanol concentrations equal to or higher than 9% shows that *Saccharomyces cerevicea* cannot breathe with such ethanol concentrations what is justified by a significant death rate of the Strain. Generally, the inhibition of the

Table 4. Effect of the exogenic ethanol concentration on the percentage of survival of the Strains of fermentation

Ethanol exogene	Kluyveromyces marxianus	Candida inconspicua	Candida xylopsoci	Saccharomyces cerevisiae	Aspergillus niger
0%	100	100	100	100	100
3%	99	95	95	100	99
4,5%	93	90	90	96	95
5%	58	50	49	63	60
5,5%	30	29	23	40	39
6%	17	15	11	22	20
7%	0	0	0	13	7
8%	0	0	0	4	0
9%	0	0	0	1	0
10%	0	0	0	0	0
11%	0	0	0	0	0

Table 5. Variation of the productivity of ethanol according to the temperature

Tmp (°C)	Kluyveromyces marxianus			Candida inconspicua			Candida xylopsoci			Saccharomyces cerevisiae			Aspergillus niger		
	Et	Bv	Pr	Et	Bv	Pr	Et	Bv	Pr	Et	Bv	Pr	Et	Bv	Pr
22	1.1	14	0.6	1	13	0.6	1.1	14	0.6	3.2	33	1.8	3.2	33	1.8
30	2.2	30	1.3	3.2	33	1.8	4.1	42	1.7	7.2	74	2.4	7.0	73	2.4
38	3.2	33	1.8	5.3	54	1.97	5.0	51	1.9	5.1	53	1.9	6.8	71	2.3
42	5.13	53	1.9	5.0	51	1.9	5.4	54	1.9	4.1	42	1.7	3.2	33	1.8
48	6.1	65	1.9	3.1	30	1.8	3.2	33	1.8	2.2	30	1.3	2.2	30	1.3
58	1.1	14	0.6	1.1	14	0.6	1	13	0.6	1.1	14	0.6	1	13	0.6

Tmp: Variation in the temperature (C); **Et**: Ethanol (%); **Bv**: Bioconversion (%); **Pr**: Productivity (g/l/h)

growth of the microorganisms is a character related to a significant number of gene for each Strain, to the adaptation and the resistance, which are acquired by the Strains in their mediums of growth [20–23].

- Determination of the optimal temperature of the production of ethanol

We studied the influence of 5 different temperatures 22 °C, 30 °C, 38 °C, 42 °C, 48 °C and 58 °C for the production of ethanol after fermentation by the various Strains used. The pH was adjusted to 5 with tartaric acid 1N, the initial concentration of fermentable sugar was 15%, the initial concentration of the inoculum was 109 per milliliter of inoculum. The production of ethanol was proportioned after 48 h of fermentation. Table 5 represents the results. According to the results of Table 5, the percentage of production of ethanol increases when the temperature at 30 °C for *Saccharomyces cerevicea* and *Aspergillus niger* with a high bioethanol production (respectively 7,2% and 7%) and we noticed a stability of bioethanol production in the interval 28 °C–34 °C, which is a margin of optimal temperature supporting the energy need necessary in order to hold the maximum growth.

In addition, for *Kluyveromyces marxianus* the optimal temperature is of 48 °C with a production of bioethanol equal to 6, 1%, then the optimal temperature for the maximum production of bioethanol by *Candida inconspicua* and *Candida xylopsoci* is between 38 °C with 42 °C (respectively 5.3% and 5.4%). In conclusion, the maximum production of bioethanol by the Strains used is according to the temperature of incubation. The results found in this part agree with other studies in literature [24, 25].

- Determination of optimal pH of bioethanol production

pH tested were between 2 and 7, the pH was adjusted by the acid tartaric 1N and NaOH 2M. The principal results found for the effect of pH on the maximum production out of bioethanol are summarized and noted in Table 6.

Table 6. Optimal pH of fermentation by the Strains used for the maximum production out of ethanol

	Optimal pH	Maximum production out of bioethanol (%)
Kluyveromyces marxianus	5.5 at 45 °C	6.1
Candida inconspicua	4.5 at 38 °C	5.3
Candida xylopsoci	4 at 42 °C	5.4
Saccharomyces cerevisiae	4.9 at 30 °C	7.2
Aspergillus niger	5.9 at 30 °C	7.0

The results obtained in this table show that at acidic pH *Candida inconspicua*, *Candida xylopsoci* and *Saccharomyces cerevisiae* give the best productions out of bioethanol. While *Aspergillus niger* and *Kluyveromyces marxianus* fermented lactose at pH similar of the dairy wastes (lactoserum and white water) which is equal to 6.5. At low pH, the percentage of production of ethanol is weak for all the tested Strains fermentation used. The reduction in pH causes a strong diffusion of protons to the cytoplasm of the cell, thereafter cellular energy necessary for the growth is used to maintain the pH intracellular constant. The found values of pH are also in agreement with several library searches [26–28].

- Effect of the type of substrate on the production of bioethanol

In this part, we studied the effect of the initial substrate of fermentation on the production of bioethanol by the Strains tested. The temperature and the pH were optimized for each Strain and the production of ethanol is proportioned after 48 h of fermentation. Table 7 summaries of the results.

The maximum production of bioethanol of each Strain is noticed under the optimal conditions of fermentation (pH and temperature) and on fermentable simple sugars (galactose and glucose). The physico-chemical or biological processes of treatment of the dairy waste increase in the medium the easily fermentable sugar rate monomeric for the production of bioethanol parallel, this production was decreased while being based on a made up substrate like lactose or on the complex lactoserum and white water for yeasts tested as for *Aspergillus niger*.

Table 7. Effect of the initial substrate on the maximum production out of ethanol by various Strains of fermentation

	Substrate (15% Fermentable sugars)	P_mEt (%)	Bv (%)	Pr (g/l/h)
Kluyveromyces marxianus	Lactoserum + white water	2.85	31	1.5
	Lactose	5.00	51	1.9
	Galactose + glucose	6.1	65	1.99
Candida inconspicua	Lactoserum + white water	0.80	11	0.6
	Lactose	4.02	41	1.73
	Galactose + glucose	5.30	53	1.98
Candida xylopsoci	Lactoserum + white water	-	-	-
	Lactose	4.12	42	1.73
	Galactose + glucose	5.40	54	1.98
Saccharomyces cerevisiae	Lactoserum + white water	-	-	-
	Lactose	3.10	34	1.85
	Galactose + glucose	7.20	74	2.44
Aspergillus niger	Lactoserum + white water	3.90	36	1.9
	Lactose	6.40	63	2.59
	Galactose + glucose	7.00	73	2.99

PmEt: Maximum production out of ethanol (%); **Bv**: Bioconversion (%); **Pr**: Productivity (gr/l/h); -: not detected.

The lactose is a disaccharide formed by galactose and classic microorganisms used in fermentation cannot carry out glucose, its conversion into ethanol. Indeed, only some yeasts belonging to *Saccharomyces* and *Kluyveromyces* or some bacteria are able to ferment lactose to ethanol directly [29–31].

In our work the lactose was fermented directly by all the Strains tested but its conversion is very easy by *Kluyveromyces marxianus* and *Aspergillus niger* with a maximum production of bioethanol for the mushroom *Aspergillus niger* (6.40%), followed by the yeast *Kluyveromyces marxianus* (5.00%). The physico-chemical and biological characteristics of the lactoserum and white water are very varied (rock salt, proteins, suspended matter, nitrogen species, phosphorated matters, chlorides and heavy metals). Several studies showed the toxic effects of these compounds on the process of alcoholic fermentation [32–35] and consequently, they inhibit the growth of the Strains used for the production of bioethanol. In addition, the difficulty, which arises, in this kind of work is to find the micro-organism able to grow under these conditions (Fig. 3) and to degrade monomeric sugars of the lactoserum without treatment for a maximum production of ethanol (3.9% for *Aspergillus niger* and 2.85% for *Kluyveromyces marxianus*). Thus, these two Strains are promising and able to treat the dairy wastes. To improve the production of bioethanol, we took into account the results of this paragraph to study the effect of *Aspergillus niger* and *Kluyveromyces marxianus* immobilization.

3.5   Kinetics of Degradation of Glucidic Substrates by the Free Cells

- Degradation by the free cells

The curves of glucidic consumption of substrate (lactose, glucose + galactose) and of production of ethanol and single cell protein biomass according to the time of fermentation by the free cells of *Aspergillus niger* and *Kluyveromyces marxianus* are represented in Figs. 3 and 4. Fermentation was carried out under the optimum conditions of pH, temperature, of agitation, the initial glucidic substrate was (15% of fermentable sugars) and 10^9 of initial inoculum and knowing that the lactose results from the lactoserum ultrafiltration (third part).

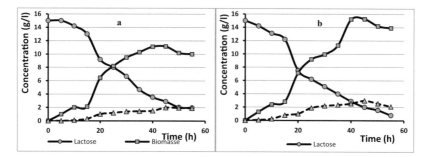

Fig. 3. Consumption of lactose and production of bioethanol and biomass by the free cells of *Kluyveromyces marxianus* (a) and *Aspergillus niger* (b)

Figure 4 presents two phases (diauxic Phenomenon). During, the first phase the bacterial cells use only glucose. After 22 h of fermentation (second phase) galactose starts to be used with a low speed of 0.1 kgS/m^{3}h^1 (Table 8). We noticed that Kluyveromyces *marxianus* and *Aspergillus niger* could ferment the totality of the glucidic matter of the dairy wastes.

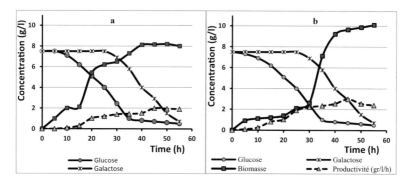

Fig. 4. Consumption of glucose and galactose and production of bioethanol and biomass by the free cells of *Kluyveromyces marxianus* (a) and *Aspergillus niger* (b)

The diauxic phenomenon in the process of fermentation can be explained during the first phase by a catabolic repression exerted by glucose on the assimilation of other sources of carbon [36]. Then, in the second phase, galactose starts to be used. We noticed that *Kluyveromyces marxianus* and *Aspergillus niger* could ferment the totality of existing free sugars. They also showed the capacity to ferment the lactose, which is regarded as a very abundant sugar in nature and less fermented by the traditional Strains used in the industry of production of ethanol. These two phases of ethanol production are confirmed by the variation of the biomass of each Strain according to time (Fig. 4).

Table 8. Kinetic parameters of the two physiological phases of fermentation of glucose and galactose by *Aspergillus niger*

	Phase I (0–22 h)	Phase II (22–50 h)
Y x/s	0.265	0.245
Y p/s	0.271	0.073
μ (H^{-1})	0.133	0.039
Qs $(g/1h^{-1})$	0.495	0.099
Qp $(g/1h^{-1})$	0.159	0.015
W	0.531	0.131
Z = (1 − W)	0.469	0.869

Yx/s: Output of production of the biomass compared to the substrate; **Yp/s:** Theoretical yield of production of the product compared to the substrate; **Y'p/s:** Experimental output of production of the product compared to the substrate; μ (H^{-1}): growth rate; **Qs:** Speed of consumption of the substrate; **Qp:** Speed of production of the product; **W:** Fraction of substrate used in the production of ethanol; $Z = 1 − W$ Fraction of the substrate used in the growth and cellular maintenance.

During the fermentation of glucidic substrates (glucose and galactose) by *Aspergillus niger* for the production of bioethanol, we obtained 54% (= W) substrate used in the cell multiplication and 46% (= Z) for the cellular maintenance in the first phase. While in the second phase, 13% are employed for the cell multiplication and 87% for the maintenance of the cellular metabolism. These results reflect the increase in the output of ethanol in the first phase and the significant fall of the output of conversion to ethanol of galactose in the second phase. Consequently, most of galactose is useful for the growth and cellular maintenance in the late phase.

Physiologically, in addition to what is explained previously, the ethanol produced during the first phase exerts a toxicity on *Aspergillus niger*, by deteriorating its plasmic membrane. Indeed, it is shown that the ethanol increases the permeability of the membrane to the protons [37], which decreases the gradient of H^+ necessary to the synthesis of the ATP. Thus, energy used for the cell multiplication is diverted to maintain the gradient of intracellular pH and consequently with the reduction in the output of ethanol during this physiological phase. This obstacle let to suggest a modification of the physiological conditions of *Aspergillus niger* to direct the metabolism towards a high production of ethanol in the second phase.

4 Conclusion

This work was interested in total valorization of dairy wastes LVRS for the renewable energy production (bioethanol).

We tested the production of bioethanol starting from lactoserum by various Strains (*Kluyveromyces marxianus, Candida inconspicua, Candida xylopsoci, Saccharomyces cerevisiae, Aspergillus niger*). The results show a variation of the resistance of the Strains tested to the various concentrations of exogenic ethanol. The best results were obtained by *Saccharomyces cerevisea* and *Aspergillus niger*.

Aspergillus niger and *Kluyveromyces marxianus* showed the capacity to grow and degrade lactose under inhibiting conditions of the fermentation and for a maximum production of ethanol (3.90% for *Aspergillus niger* and 2.85% for *Kluyveromyces marxianus*). Still, results obtained with *Aspergillus niger* encourage its use to decrease the effect polluting of LVRS.

The kinetics of degradation of glucidic substrates by the free cells of *Aspergillus niger* and *Kluyveromyces marxianus* presented the phenomenon of diauxic. In the process of fermentation, the cells prefer the use of glucose followed by the use of galactose. It was concluded whereas these two Strains in a state free could ferment the totality of the glucidic matter of the dairy rejections.

The process of valorization by the renewable energy production starting from the industrial waste seems to be encouraging and inexpensive and can be largely applied to decrease the effect on the environment.

References

1. Zahedifar, M.: Novel uses of Lignin and Hemicellulosic Sugars from Acidhydrolysed Lignocellulosic Materials. Thesis submitted for the degree of Doctor of Philosophy in the University of Aberdeen. University of Tehran (1996)
2. Prévot, H.: La récupération de l'énergie issue du traitement des déchets. Rapport d'Ingénieur général des mines. Conseil général des Mines Ministère de l'Economie, des finances et de l'industrie, Canada (2000)
3. Abidi, N.: Valorisation du lactose et du lactosérum en acide succinique par fermentation bactérienne. Doctorat en Microbiologie agricole. Université Laval, Québec (2009)
4. SGIC: Secrétariat du Groupe d'experts intergouvernemental sur l'évolution du climat. Les émissions de gaz à effet de serre s'accélèrent malgré les efforts de réduction. Communiqué de Presse du GIEC, Genève, Suisse (2014)
5. Gana, S., Touzi, A.: Valorisation du lactosérum par la production de levures lactiques avec les procédés de fermentation discontinue et continue. Rev. Energ. Ren. **1**, 51–58 (2001)
6. Jinjarak, S., Olabi, A., Flores, R.J., Sodini, I., Walker, J.H.: Sensory evaluation of lactoserum and sweet cream buttermilk. J. Dairy Sci. **89**, 2441–2450 (2006)

7. Marwaha, S.S., Kennedy, J.F., Khanna, P.K., Tewori, H.K., Redhu, A.: Comparative investigation on the physiological parameters of free and immobilized yeast for effective treatment of Dairy effluents. Physiology of immobilized cells. In: Proceeding of an International Symposium Held at Wageningen. Elsevier Science Publishers. B. V. Amsterdam (1988). Printed in Netherlands

8. Talabardon, A.M.: Acetic acid productions from milk permeate in anaerobic thermophilic fermentation. Thèse pour l'obtention du grade de docteur ès sciences techniques, Genève, Suisse (1999)

9. Bertrand, M.: Suivi de l'ATP et des protéines du biofilm dans un bioréacteur à lit fluidisé fermentant un perméat de lactosérum reconstitué. Mémoire présenté à l'Université du Québec à Chicoutimi, Canada (2002)

10. Glutz, F.N.: Fuel bioethanol production from lactoserum permeate. Doctorat universitaire Es Sciences. Ecole Polytechnique. Fédérale de Lausanne, Suisse, N° 4372 (2009)

11. Allouache, A., Aziza, M.A., Zaid, T.A.: Analyse de cycle de vie du bioéthanol. Revue des Energies Renouvelables, vol. 16, no. 2, pp. 357–364 (2013)

12. Lefrileux, Y.: Gestion des effluents d'Elevage et de Fromagerie chez les laitiers et les fromagers. Institut de l'Elevage Lefrileux. 4èmes Journées Techniques Caprines (2013)

13. Chauprade, A., Talimi, J.P.: Le changement climatique: causes et avis partagés GÉOPOLITIQUE: *Carnets du Temps*, no. 55, pp. 12–14 (2014)

14. Dubois, M., Gilles, K.A., Hamilton, J.K., Rebers, P.A., Smith, F.: Colorimetric method for determination of sugars and related substances. Anal. Chem. **28**, 350–356 (1956)

15. Gold, R.S., Meagher, M.M., Tong, S., Hutkins, R.W., Conway, T.: Cloning and expression of the Zymomonas mobilis "production of ethanol" genes in *Lactobacillus casei*. Curr. Microbiol. **33**, 256–260 (1996)

16. Yazawa, H., Iwahashi, H., Uemura, H.: Disruption of URA7 and GAL6 improves the ethanol tolerance and fermentation capacity of *Saccharomyces cerevisiae*. Yeast **24**(7), 551–560 (2007)

17. Coté, A., Brown, W.A., Cameron, D., Walsum, G.P.: Hydrolysis of lactose in lactoserum permeate for subsequent fermentation to ethanol. Am. J. Dairy Sci. **87**, 1608–1620 (2004)

18. Carlson, R., Nordahl, A.: Exploring organic synthetic, experimental procedures. Top. Curr. Chem. **166**, 1–64 (1993)

19. Olsson, L., Hahn, H.B.: Fermentation of lignocellulosic hydrolysates for ethanol production. Enzyme Microb. Technol. **18**, 312–331 (1996)

20. Eloutassi, N., Louasté, B., Chaouch, M.: Production des sucres simples à partir du lactose issu du lactosérum. Rev. Microbiol. Ind. San et Environn. **5**(2), 39–53 (2011)

21. Zhao, X., Xue, C., Ge, X., Yuan, W., Wang, Y., Bai, W.: Impact of zinc supplementation on the improvement of ethanol tolerance and yield of self-flocculating yeast in continuous ethanol fermentation. J. Biotechnol. **139**(1), 55–60 (2009). View at Publisher · View at Google Scholar · View at Scopus

22. Gomes, L.H., Duarte, K.M.R., Lira, S.P.: Increase on ethanol production by blocking the ADH2 gene expression in GFP3-transformed *Saccharomyces cerevisiae*. Greener J. Biol. Sci. **3**(1), 058–060 (2013)

23. Furlan, S.A., Schneider, A.L.S., Merkle, R., Carvalho-Jonas, M.F., Jonas, R.: Optimization of pH, temperature and inoculum ratio for the production of β-D-Galactosidase by *Kluyveromyces marxianus* using a lactose-free medium. Acta Biotechnol. **21**(1), 57–64 (2001)

24. Arroyo-Lopez, N.F., Orlić, S., Querol, A., Barrio, E.: Effects of temperature, pH and sugar concentration on the growth parameters of Saccharomyces cerevisiae, S. kudriavzevii and their interspecific hybrid. Int. J. Food Microbiol. **131**, 120–127 (2009)

25. Hadiyanto, H., Ariyantia, D., Puspita, A.A., Pinundia, D.S.: Optimization of ethanol production from lactoserum through fed-batch fermentation using Kluyveromyces marxianus. Energy Proc. **47**, 108–112 (2014)

26. Carina, S.A., Guimarães, P.M.R., Teixeira, J.A., Lucília, D.: Production of bioethanol from concentrated cheese lactoserum lactose using flocculent *Saccharomyces cerevisiae*. Poster Session: S7 – Industrial and Food Microbiology and Biotechnology: Book of Abstracts of MicroBiotec 09, 28 (2009)

27. Christensen, K., Andresen, R., Tandskov, I., Norddahl, B., Preez, J.H.: Production of bioethanol from organic lactoserum using Kluyveromyces marxianus. J. Ind. Microbiol. Biotechnol. **38**, 283–289 (2011)

28. Rodrigues, N., Rocha, A.F., Barros, M.A., Fischer, J., Filho, U.C., Cardoso, V.L.: Ethanol production from agroindustrial biomass using a crude enzyme complex produced by *Aspergillus niger*. Renew. Energy **57**, 432–435 (2013)

29. Staniszewski, M., Kujawski, W., Lewandowska, M.: Ethanol production from lactoserum in bioreactor with co-immobilized enzyme and yeast cells followed by pervaporative recovery of product – Kinetic model predictions. J. Food Eng. **82**, 618–625 (2007)

30. Agustriyanto, R., Fatmawati, A.: World Academy of Science, Engineering and Technology. Model of Continuous Cheese Lactoserum Fermentation by *Candida Pseudotropicalis*, vol. 57, pp. 213–217 (2009)

31. Ghanadzadeh, H., Ghorbanpour, N.: Optimization of ethanol production from cheese lactoserum fermentation in a batch-airlift bioreactor. J. Bioeng. Biomed. Sci. **2**, 111 (2012). https://doi.org/10.4172/2155-9538.1000111

32. Gabardo, S., Rech, R., Augusto, C., Záchia, M.A.: Dynamics of ethanol production from lactoserum and lactoserum permeate by immobilized strains of Kluyveromyces marxianus in batch and continuous bioreactors. Renew. Energy **69**, 89e96 (2014)

33. Errachidi, F., El Outassi, N., Remmal, A.: Amélioration d'une Strain de Candida tropicalis EF14 pour la production d'éthanol et l'assimilation des phénols. Thèse de Troisième Cycle en Biologie, Option Biotechnologie. Université Sidi Mohammed Ben Abdellah, Faculté des Sciences Dhar El Mahraz, Fès (1997)

34. Becerra, M., Baroli, B., Fadda, A.M., Méndez, J., Siso, M.I.: Lactose bioconversion by calcium-alginate immobilization of Kluyveromyces lactis cells. Enzyme Microb. Technol. **29**, 506–512 (2001)

35. Cotè, A., Brown, W.A., Cameron, D., Walsum, J.P.V.: Hydrolysis of lactose in lactoserum permeate for subsequent fermentation to ethanol. J. Dairy Sci. **87**, 1608–1620 (2004)

36. Ansaria, S.A., Qayyum, H.: Lactose hydrolysis from milk/lactoserum in batch and continuous processes by concanavalin A-Celite 545 immobilized Aspergillus oryzae _ galactosidase. Food Bioprod. Process. **90**, 351–359 (1985)

37. Osman, Y.A., Ingram, L.O.: Mechanism of ethanol inhibition of fermentation in Zymonas mobilis. J. Bacteriol. **164**, 173–180 (1985)

Study of Wastewater Treatment's Scenarios of the Faculty of Sciences - Ain Chock, Casablanca

Proposal I: Vertical Flow Filter

Nihad Chakri[1(✉)], Btissam El Amrani[1], Faouzi Berrada[2], and Fouad Amraoui[2]

[1] Treatment and Valorisation of Water Laboratory of Geosciences Applied to the Arrangement Engineering (G.A.A.E.), Faculty of Sciences - Aïn Chock, Hassan II University, Casablanca, Morocco
n.chakri2009@gmail.com
[2] Hydrosciences, Laboratory of Geosciences Applied to Arrangement Engineering (G.A.A.E.), Faculty of Sciences - Aïn Chock, Hassan II University, Casablanca, Morocco

Abstract. In the context of water stress in our country, alternative solutions for better management of water resources are to be developed. It's in this spirit that our research team initiated a project to design a prototype for the purification of part of the wastewater from our Faculty in order to reuse it.

This prototype is an open-air educational tool for students from different disciplines of our Faculty.

Laboratory tests for the three scenarios have been carried out to highlight the most suitable process for local conditions, namely:

Scenario I: settling basin followed by a vertical flow filter.
Scenario II: settling basin followed by two Moving Bed Biofilm Reactors.
Scenario III: settling basin followed by two Moving Bed Biofilm Reactors then a vertical flow filter.

In this study, only the wastewater treatment with vertical flow filter was addressed. So, after identification and characterization of the liquid discharges, we designed and implemented four purification prototypes when they are fed by liquid discharges, previously decanted, toilets of the Faculty.

The best reduction was obtained by combining pozzolans and reeds, which is probably due to the combined action of macrophytes, bacteria and the physical barrier constituted by the filter body.

Based on the results obtained at laboratory scale, we designed and implemented a prototype of a mini-wastewater treatment plant of the Faculty. Construction work is currently underway.

This project would allow in-depth experimental studies to define the ideal system for establishing the procedure of optimal selection to filtering material, the type of plants, the length of stay, parameters analyzed...

© Springer Nature Switzerland AG 2019
M. Ezziyyani (Ed.): AI2SD 2018, AISC 913, pp. 176–187, 2019.
https://doi.org/10.1007/978-3-030-11881-5_15

Keywords: Wastewater · Purification · Gravel · Pozzolan · Reuse · Sustainable development · Planted filter · Reeds

1 Introduction

Morocco knows a demographic expansion and an important economic evolution. This socioeconomic development affected the water resources, already limited, which reached a high level of mobilization and exploitation. Furthermore their quality becomes disturbing. Indeed, the industrial discharges are for the most of the time released into the urban effluents without preliminary treatment [1].

Conscious of this problem, the Faculty of Science Ain Chock proposes a study for the purification of a part of its liquid discharges to use it again for the cleaning and the watering of the green spaces. This project integrates several multidisciplinary units of research. It is about finalizing a compact treatment process to obtain uncluttered waste water which will be able to be used again with a low cost.

Our Faculty general strategy takes care of studies with future application. This study will allow us to master theoretical and practical knowledge regarding purge processes of wastewater applied on a real model.

They are different purge techniques of wastewater: lagooning treatment [2], activated sludge, biological disks [3]. They are interesting, but the cost, the important busy area, and the olfactive nuisances are considered as an inconvenience....

In the case of our Faculty and because of the budget and of the limited ground area [4], the most adapted purge systems will be:

1. Decantation followed by a vertical flow filter alone or combined with a biological system [5–7].
2. Decantation followed by four Moving Bed Biofilm Reactor (MBBR) process using polyurethane foam as biomedia.
3. Decantation followed by two Moving Bed Biofilm Reactor (MBBR) process followed by a vertical flow filter.

In this study, we designed and set up at the laboratory four prototypes. We use, as filtering materials, the pozzolan in two filters and the gravel in two the others. The both types of filters (pozzolan and gravel), one is planted by reeds and the second is without plantation. These filters are fed by the liquid discharge settled before and coming from the university toilet.

The objective of this study is to determine the prototypes purification ability and estimate the quality level of uncluttered waters, and the necessary conditions to finalize a purge pilot. The analyses results will be compared to the Moroccan quality standards of waters intended for the irrigation. We will check also the effect of the waste water length of time in the settling pond, and the role of the combination between macrophytes and filter materials in the purification efficiency.

2   Materials and Methods

2.1   Description of Sampling Site

The science faculty Ain chock is located southwest of Casablanca about 6 km from the Atlantic coast. The total area of its site is 3.62 ha can be subdivided into two parts. Its sewer system is the unitary type; it contains two outlets (Exit1 and Exit 2 in Fig. 1).

Fig. 1. Location of the compact wastewater treatment pilot and sampling point.

Wastewaters used in this study are coming from the student's toilet of the Faculty of Science- Ain Chock (Fig. 1).

2.2   Choice of Treatment Process

The choice of the adopted system is based, on the one hand, on extensive bibliographic studies comparing the advantages and disadvantages of different sewage treatment systems. Indeed, vertical flow filter do provide the benefits, such as:

In financial terms, this process is an inexpensive treatment technology for wastewater treatment; know that the costs for operation and maintenance are practically eliminated.

This technology is suitable for the sustainable use of locally available resources. The facility is easy and simple to build, since no complex infrastructure is needed, and there are low needs for expensive materials, while local labor can be used for the construction (pozzolan, gravel …).

Energy requirements are very low and are usually consumed for the lighting of the facility, and, possibly, for the operation of few pumps for wastewater feeding and lifting; however, pump use can be minimized or even totally eliminated with proper cascade design and exploitation of the natural ground slope to have gravity flow [8–10].

And on the other hand, this choice of this process was made on the basis of the results obtained by monitoring the evolution of the purification efficiency of the planted filters (Horizontal and vertical) set up for the treatment of wastewater from Hammam Douar Ouled Ahmed (Dar Bouaazza).

It has been proven that the efficiency of the vertical flow filter is higher than the horizontal flow filter [11].

This within the framework of a pilot project initiated by the research-action program Morocco-Germany on the Urban Farming in Casablanca (UFC) [11].

2.3   Experimental Device

The experimental prototype is constituted by 4 tubs placed in the laboratory by using the pozzolan as materials leaking out in two tubs and the gravel in the others.

For every type of substratum, a tub is crashed with Warblers Australis (reeds) and second not standing one. The diameters of the gravel and the pozzolan forming the substratum vary between 2 mm and 40 mm the top-down respectively.

The used tubs are plastical made, typical high-density polyethylene (HDPE), 0.215 m of height and 0,432 m^2 of bed area. These tubs are provided with a small valve to facilitate the treated water evacuation.

All the analysis have been done following to normalized methods Rodier and AFNOR in the Laboratory "Geosciences Applied to the Arrangement Engineering" (GAAE) by the team working on Waters Treatment and valorisation.

The physic-chemical analysis protocol used is based on the following methods:

The suspension material (TSS) is determined according to the standard T90-105 AFNOR,

The biochemical oxygen demand (BOD_5) is determined according to the instrumental method Rodier,

The chemical oxygen demand (COD) is determined by Oxidation by excess of dichromate of potassium in sulphuric medium to boiling (according to the NF standard T 90-101).

As a first step, we followed the abatement rate of the three major indicators of organic pollution (COD, BOD_5 and TSS); to see if the treatment is significant. The monitoring of total phosphor and total nitrogen, metals and bacteriological parameters is being processed to have a global vision on the efficiency of the process.

2.4   Process Description for Monitoring the Prototype of Purification

The treatment consists of a pre-treatment (screening), a primary treatment (settling) and another secondary sector (prototype in laboratory) (Fig. 2).

To measure the purification efficiency, several trials for the various stages of treatment have been done. Besides, we characterized wastewater taken to define in which category are situated the liquid discharges of our faculty.

A kinetic study of every stage of the treatment has been done:

- Trials done during the settling, the liquid discharges at the 0 h, 24 h and 48 h to determine the role of the settling.
- Trials done during the secondary treatment of the discharges taken without settling, after 24 h and 48 h of settling.

Fig. 2. Laboratory scale prototypes for wastewater treatment.

3   Results and Discussion

3.1   Characterization of Wastewater

We notice that all the physic-chemical parameters tested are in the usual range of the urban discharges in Morocco (Table 1).

Table 1. Average faculty wastewater concentrations in relation to the Moroccan standards.

Physic-chemical parameters	Unit	Average	Usual range of urban raw water [12]	SLV* of domestic discharge [13]
PO_4^{3-}	mg/L	3	–	–
$N-NH_4^+$		45,45	20–80	–
$N-NO_2^-$		0,73	<1	–
$N-NO_3^-$		0,01	<1	–
COD		541	1000	600
BOD_5		285	200–400	300
TSS		415	250–500	250

*SLV: specific limit value

The COD and the BOD_5 are parameters which allow estimating the quantity of the organic matter in the water;

- The values of the COD and the BOD_5 are below the specific limit values of domestic liquid waste [13].
- The ratio COD/BOD_5 of samples is about 1.9. Let us note that for the domestic effluents the range of this ratio is between 1,9 and 2,5 [14].

Therefore, the effluents of our university are inside the category of domestic waste water according to the Moroccan standards.

3.2   Evaluation of the Purification Power of Decantation

The settling consists of a separation between the liquid and the solid elements which allowed by the gravity. The solid materials settle at the tank bottom and form "primary sludge".

The Fig. 3 shows the abatement rate (%) of three parameters; the COD, BOD_5, and TSS during the settling from 0 h to 48 h.

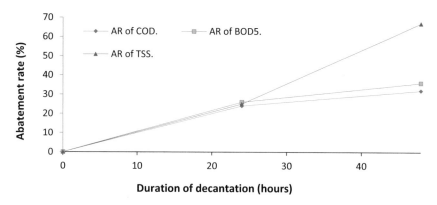

Fig. 3.   Abatement rate (AR) of COD, BOD_5 and TSS after decantation (%).

The Fig. 3 shows that:

The settling from 0 h to 12 pm: the abatement rate of COD, BOD_5, and TSS increases in the same way.

The settling from 24 h to 48 h: the abatement rate of MES is increasing quickly, while the COD and BOD_5 are increasing slightly.

The settling eliminates approximately 25% of suspension materials, COD and BOD_5 during 24 h. If we increase the settling time to 48 h, this primary treatment reduces the suspension material about 67% and about 35% the BOD_5 and the COD.

At the end of this essentially mechanical treatment, we deduce that a good part of the particles in suspension is eliminated and that the decanted waters can undergo the biological treatment.

3.3   Evaluation of the Purifying Ability of the Vertical Flow Filter

In order to evaluate the purifying ability of the prototype, we follow the three physic-chemical parameters (COD, BOD_5 and TSS).

Concentration and COD Abatement Rate

The average concentrations and the BOD_5 abatement rate of the waste water handled by the four filters are respectively illustrated by the Figs. 4 and 5.

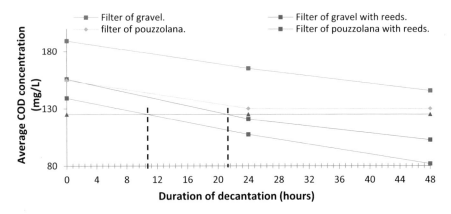

Fig. 4. Comparison of the average COD concentration in the 4 bins and the limit value.

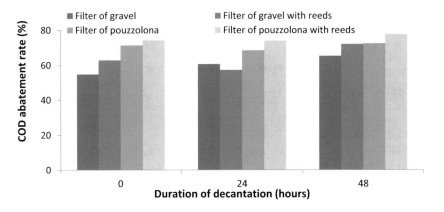

Fig. 5. COD abatement rate in the 4 bins.

The Fig. 4 shows that:

- The COD average concentration at the exit of the four filters is decreasing linearly from the 0 h till 48 h of settling, except for the pozzolan filter it decreases slightly after 24 h of settling (almost stable between 24 h and 48 h).

Both of tubs with pozzolan and the standing gravel trap reach a COD concentration lower than the limit value of waters intended for the irrigation (125 mg/l). Indeed, the standing tub of pozzolan of reeds reaches this concentration after 10 h settling. The filter with standing reed gravel reaches the same concentration later, approximately after 22 h of settling.

The Fig. 5 shows that:

- The COD abatement rate at the exit (release) of the four filters increases clearly and in the same way.

- Indeed during 48 h of settling it increases from 55% to 65% for the filter of gravel, from 63% to 72% for the filter of gravel with reeds, from 71% to 72% for the filter of pozzolan with reeds and from 74% to 78% for the filter of pozzolan.

It may be concluded that these four filters allow increasing significantly the COD dejection rate and the efficiency is higher with the filter with pozzolan crashed with reeds than the other filters.

BOD$_5$ Concentration and Abatement Rate

The BOD$_5$ average concentrations and the dejection rate of wastewater handled by the four filters are respectively showed in Figs. 6 and 7.

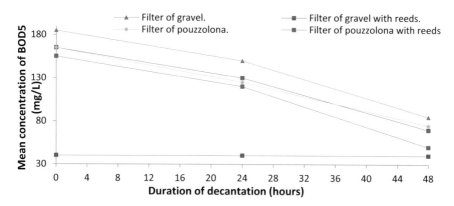

Fig. 6. Comparison of mean concentration of BOD5 in the 4 bins and the limit value.

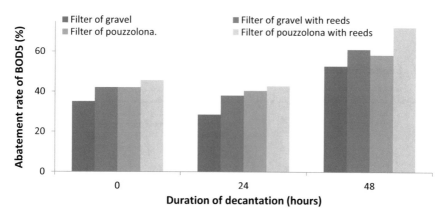

Fig. 7. BOD$_5$ abatement rate in the 4 bins.

The BOD$_5$ average concentrations at the four filters exit decreases appreciably when the settling time increases (Fig. 6).

In spite of 48 h of settling the four filters don't reach the BOD$_5$ value required for irrigation waters (40 mg/l) (Fig. 6).

Regarding the abatement rate, the Fig. 7 shows that:

The BOD_5 of dejection rate at the exit of the four filters increases clearly and in the same way. Indeed, it increases, during 48 h of settling from 35% to 53% with the filter of gravel, from 42% to 61% for the filter of gravel with reeds, from 42% to 58% for the filter of pozzolan and from 46% to 72% for the filter of pozzolan with reeds after 48 h of settling.

Then we can conclude that, reeds would have only a relative influence on the BOD_5 dejection which does not allow reaching the limit value of the waste water intended for the irrigation.

The results show also that after 48 h of settling, the BOD_5 dejection is very important with the tub for pozzolan crashed compared to the tub without reeds.

Concentration and Rate of TSS Abatement

The Figs. 8 and 9 show the average concentrations and the dejection rate of TSS wastewater handled with both of types of filters with and without reeds as well as the limit value to achieve for waters intended for the irrigation.

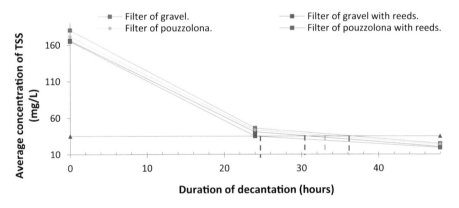

Fig. 8. Comparison of the average concentration of TSS in the 4 tanks and the limit value.

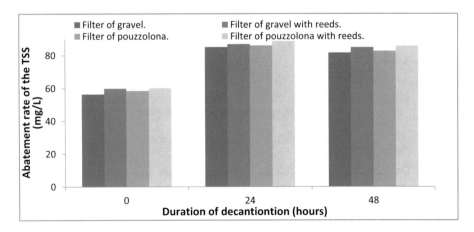

Fig. 9. TSS abatement rate in the 4 bins.

The Fig. 8 shows that:

- The average concentration of TSS at the exit of the four filters decreases, at the same speed, if we increase settling time.
- TSS at the exit of the four filters with or without reeds reaches the limit value indicated to irrigation water (35 mg/L).

Indeed, the filter with pozzolan crashed by reeds obtains this concentration after 24 h of settling, while the filter with the pozzolan not crashed reaches this concentration after 33 h of settling. Finally, this value is reached after about 30 h of settling for the filter of gravel with reeds, and after 35 h of settling for the filter with not standing gravel.

The results illustrated in the Fig. 9 show that the TSS dejection rate at the exit of four filters with or without reeds, increases clearly and in the same way. Indeed, it increases after 48 h of settling from 56% to 81% with the filter of gravel, from 60% to 85% with the filter of gravel with reeds, from 58% to 83% with the filter of pozzolan and from 60% to 86% with the filter of pozzolan with reeds. Finally we can say that reeds influence slightly the TSS dejection.

Based on these preliminary results, we concluded that the Filters of pozzolan crashed by reeds have a better purification yield than those not crashed (See Fig. 10); because:

Fig. 10. Visual comparison of wastewater before and after treatment.

The reeds have the particularity of forming a root tissue and a network of galleries that drain, provide oxygen and serve as a support for aerobic bacteria. These bacteria play a role in the degradation and mineralization of organic matter, which becomes available to plants. Thus the system doesn't produce sludge, which is composted and forms humus on site [15]. So, reeds have promoted aerobic phenomena, prevented clogging and maintained porosity sufficient to ensure the percolation of treated water [16].

The structure of the pozzolan (porosity, density, adsorption rate...) has made it possible to ensure the easy diffusion of atmospheric oxygen, and consequently the oxygen dissolved in water, a very important factor for a good purification of effluent [16].

4   Conclusion

The aim of this study is to develop an optimized process of treatment of a part of wastewater of the university of Science Ain Chock de Casablanca and to deduct the influence of the residence time of waste water in the sedimentation tank and the importance of reeds in this type of filter. According to this study we notice that: the abatement of the organic parameters of pollution (COD, BOD_5 and TSS) are satisfactory and answer the quality standards of waters intended for the irrigation.

The settling allows eliminating a big part of the suspended particles of waste water (about 67%);

Filters crashed by reeds have a better purification yield than those not crashed;

Which is probably due to the combined action of macrophytes, bacteria and the physical barrier constituted by the filter body.

The use of the pozzolan and the reeds allow having dejection rates more important than those obtained by the gravel, the COD, BOD_5 and TSS purifying efficiencies are respectively: 85%, 82% and 95% for 24 h of settling.

Based on the results obtained at laboratory scale of the various treatments' scenarios proposed, we designed and implemented a prototype of the Faculty's wastewater treatment. Construction work is currently underway (See Fig. 11).

Fig. 11. Construction work of the FSAC Wastewater Treatment Pilot.

This project would enable in-depth experimental studies, such as:

- Monitoring of the yield of several filter with different filtering materials and types of plantations.
- A MBBR performance monitoring with different biomedias and different fill rates as well to identify the oxygenation method with optimal yield.
- A highlighting of an exploitation of the different purification mechanisms.

References

1. Fagrouch, A., Berrahou, A., Halouani, H.E., Chafi, A.: Etude d'ompact des eaux usées de la ville de Taourirt sur la qualité physico-chimique des eaux de l'oued ZA. Larhyss Journal **11**, 63–77 (2012)
2. More, A., Kane, M.: Le lagunage à macrophytes, une technique permettant l'épuration des eaux usées pour son recyclage et de multiples valorisations de la biomasse. Sud Sciences and Technologies **1**, 5–16 (1998)
3. Wallace, D., Kadlec, H.: Treatment wetlands, 2nd edn. CRC Press, London (2009)
4. Alexandros, S., Christos, S.A., Vassilios, A.T.: Vertical Flow Constructed Wetlands: Eco-engineering Systems for Wastewater and Sludge Treatment. Germany (2014)
5. Saidi, A.: Approche décentralisée pour une gestion efficace des eaux usées: cas d'études dans le Grand Casablanca. Casablanca, Département de Chimie, Maroc (2015)
6. Kim, B., Gautier, M., Prost-boucle, S., Molle, P., Michel, P., Gourdon, R.: Performance evaluation of partially saturated vertical-flow constructed wetland with trickling filter and chemical precipitation for domestic and winery wastewaters treatment. Ecol. Eng. **71**, 41–47 (2014)
7. Colombano, S., Saada, A., Guerin, P.: Quelles techniques pour quels traitements - Analyse coûts- bénéfices. BRGM (2010)
8. Bruch, I., Alewell, U., Hahn, A., Hasseblach, R., Alewell, C.: Influence of soil physical parameters on removal efficiency and hydraulic conductivity of vertical flow constructed wetlands. Ecol. Eng. **68**, 124–132 (2014)
9. Mogens, H., Poul, H., Jes-la-Cour, J., Erik, A.: Wastewater Treatment, Biological and Chemical Processes. Springer, Berlin (1997)
10. Pay, D., Manzoo, R.Q., Dennis, W.: Wastewater, economic asset in an urbanizing word. Springer, London (2015)
11. Saidi, A., El Amrani, B., Amraoui, F.: Mise en place d'un filtre planté pour le traitement des eaux usées d'un Hammam et leur réutilisation dans l'irrigation d'une ferme solidaire dans le périurbain Casablancais. Mater. Environ. Sci. **5**, 2184–2190 (2014)
12. Office National d'Eau Potable (ONEP): Gamme habituel des eaux usées urbaines au Maroc (2005)
13. Arrêté conjoint no. 1607-06: Fixation des valeurs limites spécifiques de rejet domestique. Bulletin officiel No. 5448, 6–7 (2006)
14. Fatta, D., Anayiotou, S.: Medaware project for wastewater reuse in the Mediterranean countries: an innovative compact biological wastewater treatment system for promoting wastewater reclamation in Cyprus. Desalination **211**, 34–47 (2007)
15. Brix, H., Arias, A.: The use of vertical flow constructed wetlands for on-site treatment of domestic wastewater: New Danish guidelines. Ecol. Eng. **25**, 491–500 (2005)
16. Abissy, M., Mandi, L.: Utilisation des plantes aquatiques enracinées pour le traitement des eaux usées urbaines: cas du roseau. J. Water Sci. **12**, 285–315 (1999)

Optimization by Response Surface Methodology of Copper-Pillared Clay Catalysts Efficiency for the CWPO of 4-Nitrophenol

Fidâ Baragh[1,2](✉), Khalid Draoui[3], Brahim El Bali[2],
Mahfoud Agunaou[1], and Abdelhak Kherbeche[2]

[1] Laboratory of Coordination and Analytical Chemistry (LCCA),
Faculty of Sciences, Chouaib Doukkali University, Route Ben Maachou,
24000 El Jadida, Morocco
f.baragh@gmail.com, m.agunaou@gmail.com
[2] Laboratory of Catalysis, Materials and Environment (LCME),
Higher School of Technology, Sidi Mohamed Ben Abdellah University,
USMBA, EST, 30000 Fez, Morocco
b_elbali@yahoo.com, kherbecheabdelhak@gmail.com
[3] Materials and Interfacial Systems Laboratory (MSI), Faculty of Sciences,
Abdel Malek Essaadi University, Avenue de Sebta B.P. 2121, M'hannech II,
93002 Tetouan, Morocco
khdraoui@gmail.com

Abstract. The properties of copper-based pillared clays (Cu-PILBen) have been studied and compared with those of Aluminum-based clays (Al-PILBen) in the catalytic wet hydrogen peroxide oxidation (CWPO) of model phenolic compound 4-nitrophenol (4-NP) without pH adjustment. The parameters like temperature (40–60 °C), peroxide dosage (8–12 mM) and initial 4-NP concentration (50–100 mg/L) were optimized using a three-factor Box–Behnken Design (BBD) of response surface methodology (RSM). The results of this study showed that more than 90% of 4-NP was experimentally degraded using Cu-PILBen after 4 h of reaction time under optimum conditions of temperature and initial concentrations of H_2O_2 and 4-NP, which was in a good agreement with the BBD model's prediction of a 97% maximum degradation at 52 °C, initial 4-NP concentration of 50 mg/L and peroxide dosage of 10 mM.

Keywords: 4-nitrophenol · Pillared clay catalysts ·
Catalytic wet peroxide oxidation · Response surface methodology

1 Introduction

The United Nations estimates an alarming amount of 1500 km^3 of wastewater produced annually, which is six times more than the quantity of water existing in all the rivers of the world [1]. In low-income countries, only 28% of wastewater is treated and less than 8% of industrial and municipal wastewater undergoes treatment of any kind

© Springer Nature Switzerland AG 2019
M. Ezziyyani (Ed.): AI2SD 2018, AISC 913, pp. 188–202, 2019.
https://doi.org/10.1007/978-3-030-11881-5_16

[2]. Moreover, the poor access to clean water, sanitation, and hygiene pollution causes every year the death of approximately 361 000 children under 5 years, due to diarrhea, in addition to 270 000 newborn babies [1]. These large numbers are highly predicted to increase considering that the insufficient treatment facilities cannot keep up with the persistent increase in wastewater volumes and the complexity of hazardous industrial wastes evacuated into the ecosystem with minimal or no treatment. Hence, handling untreated wastewater and developing efficient, low cost and ecofriendly purifying technologies that can deal with the most recalcitrant water pollutants, is a priority nowadays. 4-nitrophenol (4-NP) is one of these hazardous pollutants that have been classified by the United States Environmental Protection Agency (USEPA), as one of the 129 priority pollutants with limit of discharge less than 0.5 mg/L [3–5]. In order to abide by the USEPA standards, researchers have focused on developing efficient treatment technologies, including physical [6–8], chemical [9, 10] and biological processes [11, 12].

Oxidation processes have often been used as pre-treatments to decrease phenolic pollutant toxicity and allow biological degradation. Since 1980, wet oxidation processes have been developed and much attention was given to catalytic wet peroxide oxidation processes (CWPO). Homogeneous wet peroxide oxidation (HCWPO), based on the redox properties of dissolved transition metals (e.g. Mn, Fe, Cu, Ce), can generate hydroxyl radicals under mild reaction conditions in the presence of hydrogen peroxide and allow a complete oxidation of organic pollutants in water into CO_2 and H_2O. Although HCWPO using copper salts, showed good water purification from recalcitrant organic pollutants, but it remains with limited utility because of the restricted favorable pH range and the need for the recovery of catalysts salts before the discharge of the treated water [13]. The use of heterogeneous catalyst containing copper cation overcomes these drawbacks.

Numerous supports have been used for developing heterogeneous catalysts. Among others: zeolites [14], polymer supported metal complexes [15], activated carbon supported metal oxides [16] and clays [7–21]. However, the low cost, abundance and interesting properties of strong surface acidity and large number of micropores, extended the use of clays as catalyst supports and especially pillared clay catalysts (PILCs) [22–24]. These materials are prepared by exchanging the charge compensating cations of the clays with large inorganic polycations, which form metal oxide clusters through thermal treatment and then the so called "Pillars" are created giving a remarkable rise of the inter-layers space [25, 26].

In the present study, CWPO of 4-nitrophenol using copper-pillared Local clay (Cu-PILBen) was investigated and compared with the performance of Al-PILBen. Furthermore, in order to optimize the degradation efficiency of 4-NP, response surface methodology based on Box–Behnken Design (BBD) was used. The effects of initial concentration of 4-NP (50–100 mg/L), H_2O_2 dosage (8–12 mM), and temperature (40–60 °C) were investigated.

2   Materials and Methods

2.1   Materials

A local bentonite sampled in the North-East of Morocco, with a cation exchange capacity (CEC) of 89 meq/100 g was employed as a starting material for the preparation of pillared clay catalysts PILCs.

NaOH (97%), HCl (37%, w/w), $Al(NO_3)_3 \cdot 9H_2O$ (99%) and $Cu(NO_3)_2 \cdot 3H_2O$ (99.99%) (all from Sigma–Aldrich) were used in the processes of purification, activation and pillaring. For the catalytic runs, 4-NP was from Merck Millipore (Germany) while hydrogen peroxide (H_2O_2) solution (30%, w/w) from Sigma-Aldrich. All the above materials were used without further purification.

2.2   Catalysts Preparation

The procedure of pillaring PILCs began firstly with the preparation of the support and then the intercalation of transition metal precursors.

Starting Material. In order to prepare the catalysts support, the purification of local clay (denoted as Raw-Ben) was done following a modification to the method previously described by Belaroui et al. [27]. Raw-Ben's fraction up to 63 μm was dispersed into distilled water for 2 h under strong magnetic stirring, allowed to settle for 16 h and then segregated by particle size (fraction $\leq$ 2 μm) following the gravimetric sedimentation procedure and finally dried at 100 °C. In order to eliminate carbonates, the resulting material was treated with HCl (0.1 N), thoroughly washed with distilled water until the elimination of the excess of chloride and then dried at 60 °C [28]. The cation exchange capacity (CEC) being about 89 meq/100 g. The homoionization of the purified clay (denoted Na-Ben) was done following Ben Achma et al. Na exchange procedure [29] and resulted in an enhancement of cation exchange capacity (CEC) of Na-Ben to 137 meq/100 g. The CEC was measured using the methylene blue method [30, 31].

Catalysts Synthesis. The Al-pillared procedure was done following a modification to the method described by Khalaf et al. [32] as follows: slow addition of NaOH solution (0.225 M) to $Al(NO_3)_3$ (0.5 M) solution, under vigorous stirring at 60 °C, up to reach a hydrolysis molar ratio OH/Al of 2. The mixture was aged for 24 h, then added drop wise, under vigorous stirring, to a 2 wt% aqueous suspension of homoionized bentonite (Na-Ben) and kept at 80°C. After the complete addition of the clay suspension, the mixture was kept at 80 °C for 3 h under continuous stirring and then aged at room temperature for one night in the presence of the mother liquor. The solid fraction was recovered by centrifugation, washed thoroughly with distilled water, dried at 60 °C and finally calcined at 400°C for 3 h. The catalyst prepared will be indicated as Al5-PILBen containing 5% wt. Al.

The preparation of Cu-PILBen was carried out following a modification to the method described by Ayodele et al. [33]. The Cu-pillaring process is the same as mentioned above, except that here $Cu(NO_3)_2$ (0.5 M) was used. The catalyst prepared will be indicated as Cu5-PILBen containing 5% wt. Cu.

2.3 Catalysts and Bentonite Clay Characterization

X-ray diffraction (XRD) patterns were measured using BRUKER D2 Phaser CuKα ($\lambda = 1.54184$ Å) with a Lynxeye detector, while X-ray fluorescence spectroscopy (XRF) analysis was obtained using an AXIOS Panalytical. The Fourier transform infrared (FTIR) spectroscopy patterns were done using a Bruker Vertex 70 infrared spectrometer with a 4 cm^{-1} resolution and 400–4000 cm^{-1} scanning range. Inductively coupled plasma atomic emission spectroscopy (ICP-AES) was used to determine the percentage of metal pillared into the catalysts as well as the leaching percentage.

2.4 Catalytic Oxidation of 4-Nitrophenol

The prepared catalysts Cu5-PILBen and AL5-PILBen were evaluated as active solids in the catalytic wet peroxide oxidation of 4-NP. The experiments were carried out in a 250 ml thermostated three neck glass flask equipped with a reflux condenser, charged with 100 ml of 4-NP at the desired temperature and under vigorous stirring. The catalyst (3 g/L) is loaded once the temperature is stabilized. Before the initiation of the CWPO reaction by the addition of peroxide, the adsorption of 4-NP by the catalyst is allowed to reach equilibrium within 15 min. This step is important in order to ascertain that the degradation efficiency of 4-NP is attributed to the oxidation reaction and not to the adsorption of the clay based catalyst. It was found out that the adsorption of 4-NP was almost negligible onto the catalyst's surface. The initial pH = 5.7 of the aqueous solution of 4-NP was not adjusted during all the reaction duration (4 h), as an attempt to evaluate the catalytic performance of the PILC catalysts without pH pre-treatment. Once the hydrogen peroxide solution was injected into the reactor, 2 ml samples were taken from the reaction mixture. The samples were centrifuged, filtered by means of 0.22 μm pore size nylon filter and analyzed immediately by UV-vis spectrophotometer VR-200 at λ_{max} of 401 nm. The degradation efficiency of 4-NP was evaluated as follows:

$$De\ (\%) = \left[\frac{C_0 - C_t}{C_0}\right] \times 100 \tag{1}$$

Where C_0 is the initial concentration of 4-NP while C_t it's concentration at the time of withdrawal.

2.5 Experimental Design and Statistical Analysis

A Box-Behnken Design (BBD) with three factors and three coded levels was used for the experimental design of the 4-NP degradation using Cu5-PILBen, in order to identify the important factors among the three studied factors (temperature, peroxide dosage and 4-NP initial concentration) which influence the CWPO process.

The response variable (Y) was fitted by a quadratic polynomial regression model equation as follows:

$$y = a_0 + \sum_{i=0}^{n} a_i x_i + \sum_{k=0}^{n} a_{ii}(x_i)^2 + \sum_{i=1}^{n-1} \sum_{j=i+1}^{n} a_{ij} x_i x_j \qquad (2)$$

Where Y represents the predicted degradation efficiency of 4-NP (%), a_0 is the offset term (constant), a_i stands for the linear coefficients, a_{ii} and a_{ij} are respectively the quadratic and interaction coefficients, while x_i and x_j are the code values of the independent input variables, which are calculated by:

$$x_i = \frac{X_i - X_{i,0}}{\Delta X_i} \qquad (i = 1, 2, 3) \qquad (3)$$

Where X_i represents the real values of the independent variable, $X_{i,0}$ is the value of X_i at the center point of the design and ΔX_i is the step change. Table 1 shows the experimental ranges and levels of the independent variables tested in the BBD.

Table 1. Variable levels of BBD for the degradation of 4-NP by Heterogeneous Fenton-like reactions on Cu5-PILBen.

Variables	Factor code	Coded and real value		
		−1	0	+1
Temperature (°C)	X_1	40	50	60
Initial 4-NP concentration (mg/L)	X_2	50	75	100
H_2O_2 dosage (mM)	X_3	8	10	12

3   Results and Discussion

3.1   Characterization of Catalyst and Local Bentonite

The XRD patterns of Raw-Ben, Na-Ben and the pillared clay catalysts are shown in Fig. 1. The basal space (001) reflection peak around $2\theta = 6°$ shifts during different stages of the pillaring process. The process of Na-exchange lowered the corresponding

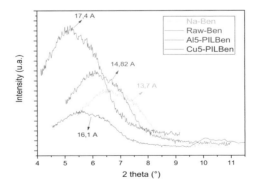

Fig. 1. X-ray diffraction profiles of Raw-Ben, Na-Ben, Al5-PILBen and Cu5-PILBen.

basal spacing, d_{001}, from 14.8 Å ($2\theta = 5.96°$) for the starting clay Raw-Ben, to 13.7 Å ($2\theta = 6.44°$) for Na-Ben. This shifting is due to the difference of the hydration diameter of Na^+, Ca^{2+} and Mg^{2+}: Na^+ (7.16 Å) < Ca^{2+} (8.24 Å) < Mg^{2+} (8.56 Å) [34, 35].

Upon pillaring with Al and Cu, the d_{001} signal shifts towards values of 17.4 Å and 16.1 Å respectively, while the rest of the structure remained unaffected. This result indicates the successful pillaring of metal contents into the activated clay.

Figure 2 represents the FT-IR spectra of Raw-Ben and Na-Ben. In the spectra of local bentonite clay, the absorption peak at 3628 cm^{-1} corresponds to the stretching vibration of structural OH group (Al-OH). The stretching vibration band of adsorbed water in the clay's sheets is clear at 3434 cm^{-1} and 1635 cm^{-1}. In fingerprint region, the bands at 1036 cm^{-1} and 477 cm^{-1} represent, respectively, the stretching vibrations of Si-O-Si and Si-O; the band at 917 cm^{-1} corresponds to Al-Al-OH bending vibration. The associated minerals are seen at 792 cm^{-1} and 1376 cm^{-1}, which implies the presence of quartz and calcite respectively; the band at 528 cm^{-1} is attributed to Si-O-Mg [34–37]. After purification the peaks attributed to these minerals are reduced in the Na-Ben, while the rest of the FT-IR spectra maintained the same allure indicating that the structure of the starting clay wasn't affected by the purification procedure and lamellar structure didn't collapse as confirmed by XRD analysis.

The ICP analysis showed in Table 2, confirms the XRD results and proves the successful pillaring-process of clay minerals by copper and Aluminum.

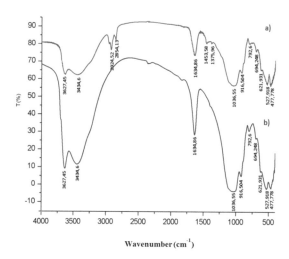

Fig. 2. FT-IR spectra of (a) Raw-Ben and (b) Na-Ben.

Table 2. Metal loading and basal spacing of PILCs samples.

Sample	Cu^a (wt.%)	Al^a (wt.%)	d_{001} (Å)
Cu5- PILBen	4.91	–	16.1
Al5- PILBen	–	4.81	17.4

^{a}Total Cu and Al content were known from ICP.

4   Design of Experiment and Response Surface Plots of CWPO of 4-NP

4.1   Statistical Analysis of the Derived Response Surface Model

Response surface methodology (RSM) using BBD was employed to statistically ana-lyze and optimize the impact of three factors [i.e. temperature, initial 4-NP concen-tration and H_2O_2 dosage] considered in this study in 4- NP degradation using copper-pillared clay. The experimental and predicted responses are assembled in Table 3.

The quality of the model was statistically evaluated by analysis of variance (ANOVA) and the results were summarized in Table 4. The model F-value of 22.3723, which is higher than critical F-value $F_{0.05, 9, 7} = 3.677$, indicates that the model is significant and there is a probability of only 0.02% that this "Model F-value" can occur due to noise [38–40]. In addition, the large values of the obtained regression coefficient R^2 (0.966) and the adjusted regression coefficient R^2_{adj} (0.923) confirm that the quad-ratic model chosen adequately predicts the removal percentage of 4-NP in the range of the parameters studied [40–42].

The polynomial equation for the 4-NP degradation efficiency (Y) obtained is as follows:

$$Y(D_e\%) = 92.876 + 0.655\,x_1 - 2.691\,x_2 - 0.157\,x_3 + 0.012\,x_1x_2 + 0.350\,x_1x_3 - 0.370\,x_2x_3 - 1.335\,x_1^2 + 1.524\,x_2^2 - 1.090\,x_3^2$$

Table 3. Box–Behnken Design and results of CWPO of 4-NP using Cu5-PILBen.

Run number	Factors			Response (D_e %)		Residual
	X_1	X_2	X_3	Observed	Predicted	
1	50	75	10	92.87	92.88	–0.01
2	60	75	12	91.50	91.30	0.20
3	60	100	10	91.00	91.04	–0.04
4	50	100	12	89.93	90.09	–0.16
5	40	50	10	95.15	95.11	0.04
6	40	75	8	90.10	90.30	–0.20
7	50	75	10	93.00	92.88	0.12
8	40	100	10	89.15	89.70	–0.55
9	50	75	10	92.86	92.87	–0.01
10	50	100	8	91.90	91.14	0.76
11	50	50	12	95.46	96.21	–0.75
12	60	50	10	96.95	96.40	0.55
13	50	50	8	95.95	95.79	0.16
14	60	75	8	90.20	90.91	–0.71
15	40	75	12	90.00	89.29	0.71
16	50	75	10	92.80	92.88	–0.08
17	50	75	10	92.86	92.88	–0.02

Table 4. ANOVA test for the response model Y.

Source	Sum of squares	Degree of freedom	Mean square	F-value	Prob > F
Model	83.991461	9	9.33238	22.3723	0.0002
Residuals	2.919979	7	0.41714		
Lack of fit	2.8979910	3	0.965997		
Pure error	0.0219885	4	0.005497		
Total	86.911441	16			

The significant of the coefficient of each independent variable evaluated and there interaction was studied based on the probability (Prob > F) shown in Table 4. Results indicate that amongst the three studied parameters, the most significant ones are temperature, initial 4-NP concentration and the quadratic effects of all three variables; while the effect of peroxide dosage and the interaction between the parameters were insignificant with Prob > F higher than 0.05.

The probability plot of residuals illustrated in Fig. 3 shows that the experimental and predicted values of the response had excellent correlation since the majority of the data points were well distributed near the straight line. The plot of the correlation of predicted response and experimental data, shown in Fig. 4, indicates that the points are randomly scattered in the range of [−0.6 − 0.6], which can confirm that the residuals are unrelated to any variable [43–45].

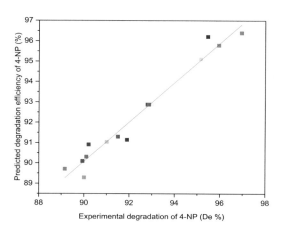

Fig. 3. Plot of predicted response versus actual response of 4-NP degradation efficiency.

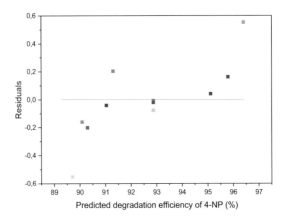

Fig. 4. Plot of residuals versus predicted response of 4-NP degradation efficiency.

4.2   Response Surface Plots of Effect of the Three Parameters Studied

Figure 5 shows the 3D response surface plots and there matching 2D contour plots corresponding to the effect of the three parameters on the degradation efficiency of 4-NP using Cu5-PILBen.

The results show that the increase of temperature from 40 to 50 °C leads to the increase of the degradation rate, from about 92% in 30 °C to 96% in 50 °C, which is in good correlation with the Arrhenius theory of the temperature's positive influence on the rate constant by enhancing the feasibility of the degradation process. However, further raise of the temperature reduces the degradation efficiency as a result of the probable thermal decomposition of peroxide at higher temperature according to Eq. 5 [33, 46].

$$C_6H_5NO_3 + 14H_2O_2 \rightarrow 6CO_2 + 16H_2O + HNO_3 \tag{4}$$

$$2H_2O_2 \rightarrow 2H_2O + O_2 \tag{5}$$

Hydrogen peroxide H_2O_2 has the same behavior as temperature on the degradation efficiency of 4-NP. In fact, the raise of peroxide dosage from 8 mM to 10 mM leads to the increase of the degradation rate. However, once the initial dosage of H_2O_2 is higher than 10 mM, the degradation efficiency drops down. This might be ascribed to the scavenging effect of excess H_2O_2 on the active hydroxyl radical $HO^{\bullet}$ as illustrated by Eqs. 6 and 7 [33, 47]:

$$H_2O_2 + HO^{\bullet} \rightarrow HO_2^{\bullet} + H_2O \tag{6}$$

$$HO_2^{\bullet} + HO^{\bullet} \rightarrow H_2O + O_2 \tag{7}$$

The 3D and 2D plots of the effect of the initial concentration of the pollutant 4-NP show that the increase of 4-NP concentration from 50 mg/L to 100 mg/L leads to a drastic decrease of the degradation efficiency from 96% to 88%. This behavior is

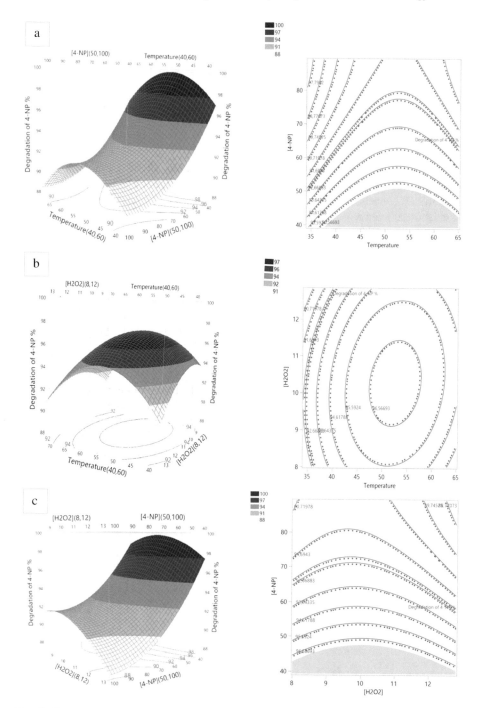

Fig. 5. Response surface plots for effect of (a) Temperature and initial concentration of 4-Nitrophenol [4-NP] (b) Temperature and peroxide dosage (H$_2$O$_2$) (c) Peroxide dosage (H$_2$O$_2$) and initial concentration of 4-NP.

probably because the reactive generated radicals HO$^•$ are insufficient to degrade rapidly a much higher concentration of organic pollutant. Moreover, the amount of intermediate compounds resulting from the degradation of 4-NP is greater in the case of higher concentration of the starting pollutant. These intermediates can be more rebellious than 4-NP itself and need a much higher concentration of peroxide to be degraded completely, which can trap the reactive radicals and forbid it from removing 4-NP [48].

4.3 Optimization and Desirability Plot

The optimum condition for the 4-NP degradation using Cu5-PILBen was obtained from the desirability plot shown in Fig. 6. Optimum conditions of temperature 52 °C, 4-NP initial concentration of 50 mM and peroxide dosage of 10 mM predicted a degradation efficiency of 97.18%.

In order to validate the optimum combination of the model, confirmatory experiment was carried out in the same condition. The experimental result showed a 97% removal using Copper based catalyst, which is in good agreement with the predicted value given by the model. Thus, the model developed by BBD coupled by RSM can be assumed as reliable and accurate.

The comparison of the catalytic performance of Cu5-PILBen and Al5-PILBen in 4-NP degradation under the optimum conditions given by the model, are presented in Fig. 7. Cu5-PILBen shows higher affinity towards the degradation of 4-NP, without pH adjustment, than Al5-PILBen which allows getting only 75% degradation efficiency. This result is justified by the strong oxidant aspect of Cu^{2+}.

Even though Al5-PILBen showed less catalytic performance than Cu5-PILBen, but it remains a good result since Al^{3+} doesn't initiate neither homolytic nor heterolytic peroxide degradation which is the essence of CWPO process [49]. This behavior is ascribed to the high amount of iron present in the starting clay, which can be more accessible upon pillaring with Keggin structure and more reactive as a result of the thermal activation at 400 °C for 3 h.

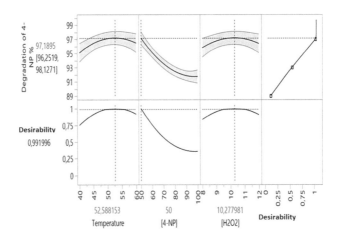

Fig. 6. Desirability plot for the degradation efficiency of 4-NP using Cu5-PILBen.

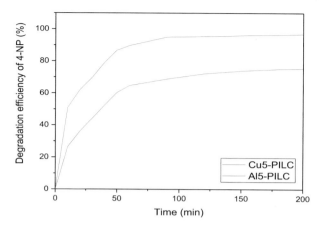

Fig. 7. Catalytic activity of Cu5-PILBen and Al5-PILBen in the 4-NP oxidation under optimum conditions, $[H_2O_2]$ = 10 mM, $m_{catalyst}$ = 3 g/L, T = 52 °C and [4-NP] = 50 mg/L.

5   Conclusions

In this study, the influence of three factors (temperature, 4-NP initial concentration and peroxide dosage) on 4-NP degradation using heterogeneous cu-pillared catalyst was investigated using Box–Behnken design coupled with response surface methodology (RSM). The quadratic model of BBD shows the presence of a high correlation between experimental and predicted values. In fact, ANOVA analysis showing a high correlation coefficient and adjusted R^2 (R^2 = 0.966 and R^2_{adj} = 0.923) and Model F-value of 22.3723 corresponding to a P-value of 0.0002, ensures a good fitting of the model's predicted response with the experimental data.

Among the three studied factors, the most significant variable was the initial 4-NP concentration (Prob > F value much less than 0.0001). In order to optimize the degradation efficiency, the desirability function was used and the BBD model developed predicted a maximum degradation efficiency of 4-NP of 97.18% under optimum conditions of 52 °C temperature, 4-NP initial concentration of 50 mM and peroxide dosage of 10 mM. Experimental validation of the model under the predicted optimum conditions showed good reliability since 97% of 4-NP was degraded experimentally under the above conditions using Cu-pillared catalyst.

Moreover, the catalytic performance of Cu-5PILBen was compared to the one of AL5-PILBen and the latter showed lower degradation of only 75%.

Acknowledgment. The authors gratefully acknowledge the financial support of CNRST-Maroc (Projets dans les domaines Prioritaires de la Recherche scientifique et du développement technologique PPR2).

References

1. World Health Organization (WHO) (2017). http://www.who.int/
2. Satoa, T., Qadir, M., Yamamotoe, S., Endoe, T., Zahoor, A.: Global, regional, and country level need for data on wastewater generation, treatment, and use. Agric. Water Manag. **130**, 1–13 (2013)
3. U.S. EPA: National Pesticide Survey: 4-NitroPhenol, National Service Center for Environmental Publications (2015)
4. U.S. Department of Health and Human Services (HSS): toxicological profile for chlorophenols. Sciences International, inc., Research Triangle Park, nc (1999)
5. Agency for Toxic Substances and Disease Registry (ATSDR): toxicological profile for nitrophenols: 2-nitrophenols and 4-nitrophenols, Public Agency for Toxic Substances and Diseases Registry, Health Service (1992)
6. Huong, P.-T., Lee, B.-K., Kim, J., Lee, C.-H.: Nitrophenols removal from aqueous medium using fenano mesoporous zeolite. Mater. Des. **101**, 210–217 (2016)
7. Zhang, J., Wu, C., Jia, A., Hu, B.: Kinetics, equilibrium and thermodynamics of the sorption of p-nitrophenol on two variable charge soils of Southern China. Appl. Surf. Sci. **298**, 95–101 (2014)
8. Hamidouche, S., Bouras, O., Zermane, F., Cheknane, B., Houari, M., Debord, J., Harel, M. Bollinger. J.-C., Baudu, M.: Simultaneous sorption of 4-nitrophenol and 2-nitrophenol on a hybrid geocomposite based on surfactant-modified pillared-clay and activated carbon. Chem. Eng. J. **279**, 964–972 (2015)
9. Shen, Y.-H.: Removal of phenol from water by adsorption–flocculation using organobentonite. Water Res. **36**, 1107–1114 (2002)
10. Mantzavinos, D., Psillakis, E.: Enhancement of biodegradability of industrial wastewaters by chemical oxidation pre-treatment. J. Chem. Technol. Biotechnol. **79**, 431–454 (2004)
11. Wang, J.-L., Zhao, G., Wu, L.-B.: Slurry-phase biological treatment of nitrophenol using bioaugmentation technique. Biomed. Environ. Sci. **18**, 77–81 (2005)
12. Jemaat, Z., Suárez-Ojeda, M.E., Pérez, J., Carrera, J.: Simultaneous nitritation and p-nitrophenol removal using aerobic granular biomass in a continuous airlift reactor. Bioresour. Technol. **150**, 307–313 (2013)
13. Kim, S.-C., Lee, D.-K.: Preparation of Al–Cu pillared clay catalysts for the catalytic wet oxidation of reactive dyes. Catal. Today **97**, 153–158 (2004)
14. El Gaidoumi, A., Chaouni Benabdallah, A., El Bali, B., Kherbeche, A.: Synthesis and characterization of zeolite HS using natural pyrophyllite as New Clay Source. Arab. J. Sci. Eng. **43**(1), 191–197 (2017)
15. Nath, N., Routaray, A., Das, Y., Maharana, T., Sutar, A.K.: Synthesis and structural studies of polymer supported transition metal complexes: Efficient catalysts for oxidation of phenol. Kinet. Catal. **56**(6), 718–732 (2015)
16. Rodrigues Carmen, S.D., Soares, O.S.G.P., Pinho, M.T., Pereira, M.F.R., Madeira Luis, M.: p-nitrophenol degradation by heterogeneous fenton's oxidation over activated carbon-based catalysts. Appl. Catal. B **219**, 109–122 (2017)
17. Timofeeva, M.N., Khankhasaeva, S.T., Talsi, E.P., Panchenko, V.N., Golovin, A.V., Dashinamzhilova, E.T., Tsybulya, S.V.: The effect of Fe/Cu ratio in the synthesis of mixed Fe, Cu, Al-clays used as catalysts in phenol peroxide oxidation. Appl. Catal. B **90**, 618–627 (2009)
18. Carriazo, J., Guélou, E., Barrault, J., Tatibouët, J.-M., Moreno, S.: Catalytic wet peroxide oxidation of phenol over Al–Cu or Al–Fe modified clays. Appl. Clay Sci. **22**, 303–308 (2003)

19. Carriazo, J., Guélou, E., Barrault, J., Tatibouët, J.-M., Molina, R., Moreno, S.: Synthesis of pillared clays containing Al, Al-Fe or Al-Ce-Fe from a bentonite: characterization and catalytic activity. Water Res. **39**, 3891–3899 (2005)

20. El Gaidoumi, A., Loqman, A., Chaouni Benadallah, A., El Bali, B., Kherbeche, A.: Co(II)-pyrophyllite as catalyst for phenol oxidative degradation: optimization study using response surface methodology. Waste Biomass Valor (2017). http://dx.doi.org/10.1007/s12649-017-0117-5

21. Sotelo, J., Ovejero, G., Martinez, F., Melero, J., Milieni, A.: Catalytic wet peroxide oxidation of phenolic solutions over a $LaTi_{1-x}Cu_xO_3$ perovskite catalyst. Appl. Catal. B: Environ. **47**, 281–294 (2004)

22. Pinnavaia, T.J.: Intercalated clay catalysts. Science **220**, 365–371 (1983)

23. Gil, A., Gandía, L.M., Vicente, M.A.: Recent advances in the synthesis and catalytic applications of pillared clays. Catal. Rev. Sci. Eng. **42**, 145–212 (2000)

24. Gil, A., Korili, S.A., Vicente, M.A.: Recent advances in the control and characterization of the porous structure of pillared clay catalysts. Catal. Rev. Sci. Eng. **50**, 153–221 (2008)

25. Vicente, M.A., Gil, A., Bergaya, F.: Pillared clays and clay minerals. In: Bergaya, F., Lagaly, G. (eds.) Handbook of Clay Science, 2nd edn. Elsevier, Amsterdam (2013)

26. Zhu, J., Wen, K., Zhang, P., Wang, Y., Ma, L., Xi, Y., Zhu, R., Liu, H., He, H.: Keggin-Al30 pillared montmorillonite. Microporous Mesoporous Mater. **242**, 256–263 (2017)

27. Belaroui, L.S., Millet, J.M.M., Bengueddach, A.: Characterization of lalithe, a new bentonite-type Algerian clay, for intercalation and catalysts preparation. Catal. Today **89**, 279–286 (2004)

28. Bergaya, F., Lagaly, G.: Purification of natural clays. In: Bergaya, F., Lagaly, G. (eds.) Handbook of Clay Science, pp. 213–219. Elsevier, Amsterdam (2013)

29. Ben Achma, R., Ghorbel, A., Dafinov, A., Medina, F.: Copper-supported pillared clay catalysts for the wet hydrogen peroxide catalytic oxidation of model pollutant tyrosol. Appl. Catal. A **349**, 20–28 (2008)

30. Hang, P.T., Brindley, G.W.: Methylene blue adsorption by clay minerals. determination of surface areas and cation exchange capacities (clay-organic studies XVIII). Clays Clay Miner. **18**, 203–212 (1970)

31. Rytwo, G., Serben, C., Nir, S., Margulies, L.: Use of methylene blue and crystal violet for determination of exchangeable cations in montmorillonite. Clays Clay Miner. **39**(5), 551–555 (1991)

32. Khalaf, H., Bouras, O., Perrichon, V.: Synthesis and characterization of Al-pillared and cationic surfactant modified algerian bentonite. Microporous Mater. **8**, 141–150 (1997)

33. Ayodele, O.B., Hameed, B.H.: Synthesis of copper pillared bentonite ferrioxalate catalyst for degradation of 4-nitrophenol in visible light assisted fenton process. J. Ind. Eng. Chem. **19**, 966–974 (2013)

34. Wang, S.W., Dong, Y.H., He, M.L., Chen, L., Yu, X.J.: Characterization of GMZ bentonite and its application in the adsorption of Pb (II) from aqueous solutions. Appl. Clay Sci. **43**, 164–171 (2009)

35. Yuan, P., Annabi-Bergaya, F., Tao, Q., Fan, M.D., Liu, Z.W., Zhu, J.X., He, H.P., Chen, T.H.: A combined study by XRD, FTIR, TG and HRTEM on the structure of delaminated Fe-intercalated/pillared clay. J. Colloid Interface Sci. **324**, 142–149 (2008)

36. Eren, E., Afsin, B.: An investigation of Cu (II) adsorption by raw and acid-activated bentonite: a combined potentiometric, thermodynamic, XRD, IR. DTA study. J. Hazard. Mater. **151**, 682–691 (2018)

37. El Miz, M., Akichoh, H., Berraaouan, D., Salhi, S., Tahani, A.: Chemical and physical characterization of moroccan bentonite taken from nador (north of Morocco). Am. J. Chem. **7**(4), 105–112 (2017)

38. Loqman, A., El Bali, B., Lützenkirchen, J., Weidler, P.G., Kherbeche, A.: Adsorptive removal of crystal violet dye by a local clay and process optimization by response surface methodology. Appl. Water Sci. (2016) https://doi.org/10.1007/s13201-016-0509-x

39. Li, H., Li, Y., Xianga, L., Huanga, Q., Qiua, J., Zhanga, H., Sivaiah, M.V., Baronb, F., Barrault, J., Petit, S., Valange, S.: Heterogeneous photo-fenton decolorization of orange II over Al-pillared Fe-smectite: response surface approach, degradation pathway, and toxicity evaluation. J. Hazard. Mater. **287**, 32–41 (2015)

40. Hamdi, H., Namane, A., Hank, D., Hellal, A.: Coupling of photocatalysis and biological treatment for phenol degradation: application of factorial design methodology. JMES **8**(11), 3953–3961 (2017)

41. Tripathi, P., Srivastava, V.C., Kumar, A.: Optimization of an azo dye batch adsorption parameters using Box-Behnken design. Desalination **249**, 1273–1279 (2009)

42. Sharmaa, P., Singha, L., Dilbaghi, N.: Optimization of process variables for decolorization of disperse Yellow 211 by bacillus subtilis using Box-Behnken design. J. Hazard. Mater. **164**, 1024–1029 (2009)

43. Ayodele, O.B., Lim, J.K., Hameed, B.H.: Degradation of phenol in photo-Fenton process by phosphoric acid modified kaolin supported ferric-oxalate catalyst: optimization and kinetic modelling. Chem. Eng. J. **197**, 181–192 (2012)

44. Diya'uddeen, B.H., Abdul Aziz, A.R., Ashri Wan Daud, W.M.: Oxidative mineralisation of petroleum refinery effluent using Fenton-like process. Chem. Eng. Res. Des. **90**(2), 298–307 (2012)

45. Yetilmezsoy, K., Demirel, S., Vanderbei, R.J.: Response surface modeling of Pb(II) removal from aqueous solution by Pistacia vera L.: Box-Behnken experimental design. J. Hazard. Mater. **171**, 551–562 (2009)

46. Caudo, S., Genovese, C., Perathoner, S., Centi, G.: Copper-pillared clays (Cu-PILBen) for agro-food wastewater purification with H_2O_2. Microporous Mesoporous Mater. **107**, 46–57 (2008)

47. Pignatello, J., Oliveros, E., MacKay, A.: Advanced oxidation processes for organic contaminant destruction based on the fenton reaction and related chemistry. Crit. Rev. Environ. Sci. Technol. **36**, 1–84 (2006)

48. Feng, H., Le-cheng, L.: Degradation kinetics and mechanisms of phenol in photo-fenton process. J. Zhejiang Univ.: Sci. **5**, 198–205 (2004)

49. Mojović, Z., Banković, P., Milutinović-Nikolić, A., Dostanić, J., Jović-Jovičić, N., Jovanović, D.: Al, Cu-pillared clays as catalysts in environmental protection. Chem. Eng. J. **154**, 149–155 (2009)

Lean in Information Technology: Produce the Human Before the Software

Soukayna Belkadi[✉], Ilias Cherti, and Mohamed Bahaj

Department of Math and Computer Science,
Faculty of Sciences and Technology, Hassan 1st University, Settat, Morocco
soukayna.belkadi@gmail.com

Abstract. Governments and private companies are under increasing pressure to improve their efficiency in delivering more and better services to citizens. Faced with the new industrial context characterized by the opening of markets, the liberalization of trade and the emergence of information technologies (IT); this companies are forced to be more flexible to survive. As a result, IT activities are becoming increasingly important in the operation of government business. Actually the Information systems have become more complex, especially with the introduction of software packages based on fully technical platforms, this complexity and duplication of business information can lead to additional costs in terms of development and maintenance. In this context that the combination between the management of information system and lean principles namely in piloting software development projects could bring more and more fruitful results and will seeks incremental waste reduction and value enhancement.

We will try in this paper to show how Lean, from the world of manufacturing, applies to the field of information systems security can improve the performance and the security of the information system. This paper will be broken into four sections: Sect. 1 discuss the evolution of lean from the world of manufacturing to the IT functions, the second section discuss the history and the challenges of the information system, the third section presents the IT governance tools and the limits of their application and the last section recommends a methodology to improve the IT maturity of an organization by discussing how Lean can drive the success of IT environments of organizations. It ends by presenting the challenges of the application of lean thinking to the IT functions.

Keywords: Lean thinking · Information system · IT governance ·
Management of information system (MIS)

1 Introduction

Based on the Japanese economic model, Lean Management has been successfully applied by companies around the world, to ensure its continuous improvement, mainly in manufacturing functions. Recently, the digital transformation has been accelerating for 30 years and profoundly modifying our lifestyles and organizations. Companies are faced with having powerful and robust information systems.

Moreover, the information system is experiencing a real change with the arrival of New Information and Communication Technologies, This situation, coupled with the

© Springer Nature Switzerland AG 2019
M. Ezziyyani (Ed.): AI2SD 2018, AISC 913, pp. 203–213, 2019.
https://doi.org/10.1007/978-3-030-11881-5_17

globalization of economies, creates a turbulent economic environment around companies. Thereby several parameters define the level of a company in the market, not only its productivity but also the robustness and security of its information systems.

It is in this context that the interest to investigate a wider application of Lean Management especially in service functions increased.

The Lean Approach ensures in particular the satisfaction of the customer through the elimination of any type of waste, this approach is often absent from the operational optimization of the IS Directions. However, it is not clear how Lean Management can be applied to IT organizations. Therefore, the goal of this paper is to translate the concept of Lean used in production functions to the field of service functions namely to the information system. We recommend adjusting Lean to the nature of IT management. Attention will be given to the amelioration of the productivity and the security of information system by using Lean principles.

2 The Evolution of the Lean Concept

Coming from Toyota's total quality method, like many others (Six Sigma,…), the Lean method is not revolutionary, but its principles are currently being updated wherever rationalization is needed, especially in computer science. In the Lean method, two principles drive the creation of value:

- The value must be estimated from the point of view of the customer;
- The value must "flow without interruption" along the design-production-distribution chain.

The origins of the Lean thinking go back to the Japanese manufacturers. (Shingo 1981; Ohno 1988; Monden 1993). After 1990, the orientation of the workshop gradually widened, characterized by the design by many detractors. This evolution has focused on quality (early 1990s), cost and delivery (late 1990s), customer value from 2000. Toyota's philosophies have been shaped by the personalities, ethics and abilities of its creators in the Toyota family. Lean's principles are firmly grounded in scientific wisdom and methods (Bicheno and Holweg 2009) [3].

Moreover, the term "lean" has been increasingly popularized in the seminal book "The Machine Who Changed the World" (Womack, Jones and Roos, 1990), which emphasized the significant performance gap between Japanese and Western automotive industries. He described the key elements for a better industrial performance that is part of the Lean concept, as Japanese business methods used less than everything: human effort, investment, facilities, inventories and time in manufacturing, Product development, parts supply and customer relations.

In addition, the Lean concept is based on the concept of more productive and more economical management of processes and therefore of the company. It aims to satisfy customers by reducing the cost and investment of production, making the most of the resources, reducing the stock and the production cycle. Lean concept allows companies to reduce costs continuously, increase quality and shorten the supply cycle in order to meet customer expectations to the maximum and adapt to environmental conditions by eliminating all types of waste by using a lot of tools such as Kanban which is Material

Flow Control mechanism (MFC) to deliver the right quantity of parts at right time, Single Minute Exchange of Dies (SMED)/One-Touch exchange of Die (OTED) to reduce the changeover time by converting possible internal setting time (Carry out during machine stoppage) to external time (performed while the equipment is running, and the Value stream Mapping (VSM) which provides a clear and simple vision of the process [4].

Several books were inspired by the lean concept used in the industrial sector, to apply it to the information system. The beginning of Lean application in IT activities first appeared in Mary and Tom Poppendieck's book "Lean Software Development" published in 2003; they adapted manufacturing methods to software development throughout seven key principles of lean:

(1) Eliminate waste;
(2) Build integrity;
(3) See the whole;
(4) Decide as late as possible:
(5) Deliver as fast as possible;
(6) Amplify learning;
(7) Empower the team.

It is in this context that the application of lean principles to the development of information system will help to control and secure the information system while eliminating any kind of waste that creates no added value. The role of the information system in the enterprise and the benefits of applying lean in information systems will be described in the next section.

3 History and Challenges of the Management of Information Systems

3.1 Evolution of Information Systems

The concept of management of information system appeared in the 1960 s in the United States (in management department of Minnesota University). It was, then, soon adopted by many management academic centers as a modern scientific attitude. It was primarily designed for managing organizational applications. Hence the name "Management Information System" which is composed of three concepts: "management", "information" and "system". From the beginning, the main goal of MIS was to assist decision-making by presenting the scientific solutions and techniques required for the managers and decision-makers of the company to design, implement and manage automatic information systems in companies.

Indeed, the information system management is directly related to the concepts of decision support. The technical infrastructure of the systems, as the foundation of information management within the organization and the applications as business tools, quickly make it a strategic issue for companies [9].

3.2   Issues of the Information System

To meet the IT needs of a structure, the coherence and agility of the information system is primary to effectively integrate new requirements. From another point of view, it must at the same time integrate new technologies and the possibilities of data analysis, information processing and above all ease of implementation to support the processes.

Continuity of service after a disaster becomes a major issue also motivated by regulatory standards for the good health of a company. The Information Systems Department has the obligation to ensure that the technical standards and human procedures are respected during a disaster but also that the organizational management allows solving the problems quickly and efficiently. It is in this context that the security of information systems is a key issue for IT systems because the vulnerability of systems and human manipulation are essential factors that the Director of Information Systems (DSI) analyzes in depth, by asking about: the threats every company needs to pay attention to? What is the maturity of the processes put in place within an organization aimed at guaranteeing Confidentiality, Availability and Integrity of data?

The control of information systems is also one of the most important issues of the company. While computerization of enterprises could originally have been seen as a simple problem of automation of administrative tasks, the efficient and effective use of technologies has now become strategic and it concerns all organizations, whatever their size and area of activity [2].

The problem that companies are currently facing with their information system is that it has a growing number of applications running on different systems for more and more diverse users, which they use from increasingly diverse hardware platforms. In addition to these applications, the company often has a multitude of databases working with different technologies, several repositories on heterogeneous platforms. This heterogeneity is highlighted when the company concerned wishes to develop new applications by reusing as much as possible the existing components within the IS [1].

It is in this context that companies have moved from corporate governance to IT Governance, the goal is to build the decision-making system, focus alignment of business processes and setting up permanent communication. Referring to the most well-known governance tools that we will define in the next paragraph.

4   IT Governance Tools

4.1   Control Objectives for Information and Technology (COBIT)

This method was designed by the Information Systems Audit and Control Association (ISACA) about ten years ago. Which provide several tools allowing managers to bring control needs closer to technical solutions and risk management. It seeks to frame the entire informational process of the company from the creation of information to its destruction to ensure accurate quality monitoring;

The originality of CobiT is to create a link between stakeholders and CIO, which often requires a small cultural revolution for both actors of the Direction of information system in their ivory tower for trades and general management who would ignore superbly the strategic character of the IS. The key point behind this approach is the establishment of constructive dialogues at all levels of the organization, between stakeholders and CIOs [12].

The CobiT repository, with its 34 generic processes, is a proposal that can be revised to be adapted to the organization's own cartography. In the same way, CobiT can easily be linked to other market references (ISO 27001, ITIL for the Information Technology Infrastructure Library or CMMI for Capability Maturity Model Integration) by building a reference framework that meets all the requirements. This is all the more true since the processes of CobiT are sometimes global and are often interpreted as "macros" of more specialized references. CobiT is therefore a unifying framework.

4.2   Information Technology Infrastructure Library (ITIL)

Information Technology Infrastructure Library (ITIL) was created within the UK Public Trade Office, it offers a structured collection of best practices for the management of the ITIL information system, and a structured, process-oriented development framework focused on the customer. The customer is indeed the founding point and the heart of the approach, using ITIL allows companies to increase user and customer satisfaction with IT services and to improve service availability.

ITIL is the most widely adopted tool for ITSM. This practice can be adapted for use in all commercial and organizational environments. ITIL's Value Proposition Focuses on IT service provider (internal or external provider) including the objectives and priorities of a customer, and the role that IT departments in achieving these goals [10].

To sum up, ITIL improves the functions in all the entities of the company. Customers will be delighted with the improved quality of IT services through the execution of consistent and repeatable processes.

4.3   Capability Maturity Model Integration (CMMI)

Capability Maturity Model Integration (CMMI) is a model for evaluating the design process of software and allows the company to classify its practices in five levels of maturity: delay, quality, cost and reliability, its purpose is to measure the capacity of projects to be completed correctly in terms of deadlines, functionalities and budget.

CMMI helps to identify and achieve predefined business goals in the business by answering the question "how do we know?", it helps also to develop better products, get closer to the customer and meet their needs.. It is composed of a set of "process areas" that must be appropriate to the company's policy. CMMI does not define how society should behave but defines what behaviors need to be defined. In this way, CMMI is a "behavioral model" and a "process model".

These governance tools are complementary; indeed COBIT's high-level control objectives can be achieved through the implementation of ITIL best practices and can be measured and evaluated by the CMMI model.

To meet the need of developing better products and get closer to the costumer, the management of information systems was defined in several ways by several authors that we will discuss below:

MIS is a system which receives data from different units and produces information and provides managers in all levels with relative, just-in-time, precise and uniform information for decision-making (Safarzade and Mansoori 2009).

MIS is a manual or computer system which improves every organization's just-in-time use, management and processing of data and information (Feizi 2005).

MIS involves official methods of providing precise and just-in-time information to facilitate managers' decision making processes while planning, controlling and making effective and optimal decisions in the organization (Momeni 1993).

As can be seen, all definitions define MIS as a "system" with inputs and outputs whose purpose is to provide the right information at the right time, this principle is the main objective of the lean concept which aims in particular to get closer to the customer to provide the right product at the right time with the right price by eliminating as much as possible non-value-added tasks. This concept could therefore be associated with the information system where the product provided is the information that must be useful for its users.

5   Lean Concept Applied to the Information System

5.1   Mudas of the Information System

Lean Information Technology is the application of Lean manufacturing and Lean services principles to the development and management of information technology products and services. It designs the information system that could provide right information to the right people, at the right time and first times, it aims also, the elimination of waste that adds no value to a product or service by using particular principles and methods.

This method is complementary to other project management techniques like Scrum that allows teams to focus on delivering product and improved communication it has been designed for collocated software development and it focuses on project management institutions where it is difficult to plan ahead [8], unlike the Lean that was initially applied to the production systems and actually it can be transported to the information system; it allows companies to achieve permanent improvement and facilitates the work of teams, eliminate non-values added activities and reduce complexity of systems.

In general, when we consider developing a system by Lean Management Information system, we normally refer to all the tools of lean used in production system such as VSM, PCA, SMED and specially the Japanese 5S: Structurize, Systemize, Standardize, Self- Discipline, and Sanitize. These lean tools can help companies to minimize the sources of waste (Muda) in the information system defined in Table 1.

These sources of waste can be improved by using Lean practices, namely Failure Modes, Effects and Criticality Analysis (FMECA) to avoid failures due to unstable processes. Optimization packages to put into production (decoupling applications) can

Table 1. Sources of waste in the information system (MUDA)

Mudas	Examples
Defaults	• Repetitive activity;
	• Unstable processes;
	• Low level of customer relationship management;
	• No compliance with cost/quality/time commitments
Waiting time	• Poor communication between the IT and the production Departments that causes
	• A delay between service and system development processes
Overproduction	• Deliveries of unused features;
	• Bad definition of customer needs which generates a production earlier;
	• Faster or in greater than the quantity expressed by the customer (Level of service too high (24 h/24 h–7 days/7 days for an application used on working days)
The unusual use of knowledge	• Low capitalization of knowledge
	• Non-reuse of components (codes …)
	• Lack of tools and skills management process
Activities without added value	• Activities that do not add value to customers, such as presenting technical dashboards to managers
Unnecessary movement	• Emergency response to recurring problems
	• Frequent change of environments (technical bases, framework, tools …)
	• Multi staffing: too many topics managed in parallel, no prioritization
	• Multiple interventions on incidents without resolution

also help to minimize the waiting time in applications, implementation of processes and tools for skills management, implementation of capitalization tools and processes, the creation of sharing times between teams including the costumer and the implementation of collaborative tools (wiki, collaborative workspace, search engines …) can help capitalize knowledge and process review by identifying and eliminating non-value-added tasks (VSM, MIFA).

5.2 Lean Applied to the IT Governance Tools

The adapted information system is based on the agile development approach, provide information in the field, especially to middle management to ensure good coordination between the decisions that emanate from the top management and their application in the market, and not just give reporting to the central Direction. The goal is to bring the decision and the decision maker closer to the market.

Therefore, agile development approaches can rely on Lean Management. These approaches are based on cycles of iterations with, for example, frequent deliveries (customer feedback), reusable automated tests, incremental designs, collaborative work, etc. Agile design requires companies to have flexible structures, with high technical quality, as opposed to large monolithic projects.

As we mentioned in the previous paragraph, ITIL, used in the agile method as a conceptual framework that promotes foreseen tight interactions throughout the development cycle. A series of books in the reference [5, 6] recommended practices on a broad range of IT management topics.

However, ITIL has some limits and is still reactive in nature, as a response to one or more incidents. It does not define the time consumption and does not provide insight into gaining efficiencies, nor does it address the leadership issues of organizational change. Furthermore, the implementation of an ITIL initiative can be difficult since it is a top down approach.

The CMMI also, unlike Lean IT, doesn't directly address sources of waste such as a lack of alignment between business units and the IT function or unnecessary architectural complexity within a software application, increased cost and overheads energy, waiting slow application. Then, using the Value Stream in the IT provides some services by the IT function to the organization that can be used by costumer, suppliers and employees, it analyses services into their component process steps.

It is in this context that the application of Lean management in a company can be associated with the development of information systems; indeed incremental developments also follow the iterative development of quality and processes.

Lean helps to control its software delivery times, to ultimately result in a streaming delivery. It can also help an IT support department to significantly reduce the time to correct incidents, by identifying and eliminating major waste in the chain of resolution and also by setting up trainings between team members to raise in competence all the collaborators.

Lean thinking uses a lot of tools to organize work namely:

– Kanban: First invented in the automotive sector, it is also possible to adapt Kanban to software creation. It helps to start where the team is in its maturity of knowledge of processes. It doesn't need to start a cultural revolution right away, as Scrum demands; Companies can keep the same people and the same roles. It can also be a solution for teams who cannot switch to Scrum for corporate cultural reasons but still want to be more agile.
– Collective visualization: One of the keys to Lean is the development of person's skills through job training or problem solving. Lean information management attaches great importance to the collective visualization of the (performance, problems …) and on its application to the computer science, where everything tends to happen in the computer and networks.

The lean principle in IT uses the principle of producing the human before producing software. This combination between lean and information system creates two value one for the customer by facilitating access and navigation on the IS and another value for employees by facilitating the processing and access to information, it allows to devote time to activities that really create value [11].

- Key Performance Indicator (KPI): It is generally difficult to measure the value for an employee working on an information system, the customer for example when he is not satisfied with the service offered by the IS, he sends complaints, comments on the internet …, which is not the case for an employee. Computer scientists are struggling to measure the benefits provided by the various functionalities of the developed IS in a company that has a poor knowledge of its operational performance, which can lead to provide useless information; employees are forced to obey the instructions issued by the enterprise system, they are then transformed not an active agent expert, but simple "clickers" obeying instructions. In this case the computer team does not bring added value to the IS [11].
- Value stream mapping and HEIJUNKA: help to know all source of waste and eradicate non value adding processing and poor costumer services, ovoid the over production and application changes and also leveling the customer demand by using the HEIJUNKA In order to overcome the fluctuation of customer demand, Without levelling, this fluctuation leads to underutilized capacities such as man and machine and specially the business and IT misalignment.
- Continuous improvement: follow the PDCA cycle, to create value, which must be a response to a need expressed by the costumer and must flow throughout the entire logistics chain of the company, in this case, the use of lean tools can help IT professionals better understand the value of their information system, with key performance indicators to define the value by referring to the users point of view and not the computer designer.

Identifying the value sought by the customer (easy navigation, confidentiality of data, good interface….) is usually a difficult task to do, as an example when the customer wants to make purchases online, the purchase process is much simpler when the site just asks the address and the payment method of the client, while other site asks to create an account and to activate this account by clicking on a link sent by email before being able to order. This complexity makes the customer no longer recommend on the same site. Lean can help to identify the value of customer by using the grid of analysis of the value elaborated by Daniel Jones and James Womack which define three types of activities in the value stream:

- Value-Added: Those activities that unambiguously create value.
- Type One Muda: Activities that create no value but seem to be unavoidable with current technologies or production assets.
- Type Two Muda: Activities that create no value and are immediately avoidable [7].

To sum up, information system using lean tools recommend that:

- Reports and other outputs from the system should only be produced if they add value and if they are useful to the decision makers and that they should only be sent to those who need them;
- Information should be processed quickly so that users do not have to wait for it;
- Continuous improvement: the providers and users of information should meet regularly to review the usefulness of existing information and identify improvements;

- Information systems should be flexible enough to meet special ad hoc needs or changing needs of managers over time. An information system that can only produce a standard set of reports is not lean. A system that allows managers to create their own customized reports from databases is more likely to be Lean.

5.3   Challenges of the Application of Lean in IT

Lean IT Still Has Challenges for Value-Stream Visualization, Unlike lean production, which is the base of Lean IT principles, it relies on digital and intangible rather than physical and tangible streams of value, these flows can be difficult to quantify and visualize, another challenge that companies may face when they use lean in their information system is related to cultural change, computer scientists always tend to work with their technical knowledge and do not care to integrate the relationship with the profession as a major axis of development of skills.

The implementation of Lean thinking can be broken by several elements such as human conflicts within the teams, the resistance to change, the principle "think product, not project" which is very disturbing for a team of developers who have not been trained and accustomed to this. The customer who has become more and more demanding and who does not know what he wants. As result, the program becomes more complex and contains several iterations; therefore the code will be insufficiently mastered.

This presents a challenge in terms of determining the lean practices that are applicable to IT support service environment and in devising ways to integrate these lean practices with process improvement framework like CMMI.

6   Conclusion

This paper discusses the development of a new approach that was initially applied to the industrial world to support the improvement and productivity of the information system. It is argued that there are many tools and methods of IT governance to improve some aspects of IS, but none of these methods allow the company to measure the robustness of its information system or to know and eliminate the non-value added activities. Nobody can deny the great revolution achieved through the application of lean in the industrial sector. The basic proposition of this article is that this tool could also be applied to the service functions and in particular to the IS where information management can be considered the product to be improved in order to meet the needs of the end user.

But, Lean IT Still Has Challenges for its application. These problems will be the subject of a future paper, we will address the obstacles that business can encounter by introducing lean into its information systems, through a case study and we will define how to move from the methods used in the industrial world to those used in the information systems to successfully implement the lean IT management tools.

Acknowledgement. At the end of this Paper, we express our gratitude and our sincere thanks to all those who contributed directly or indirectly to the realization of this article in particular M. Cherti and M. Bahaj who assisted us during the various stages of our work. Their guidance and valuable advice and availability have allowed us to master and overcome the challenges of this communication.

References

1. Elidrissi, D., Elidrissi, A.: Contribution des systèmes d'information à la performance des organisations: le cas des banques. la Revue des Sciences de Gestion **241**(1), 55–61 (2010)
2. Laudon, K.C., Laudon, J.P.: Management Information System, Managing the Digital Firm, 12th edn, pp. 41–50. Prentice Hall, Upper Saddle River (2012)
3. Emiliani, M.L.: Origins of lean management in America: the role of Connecticut businesses. J. Manag. History **12**(2), 167–184 (2006)
4. Grzelczak, A., Werner-Lewandowska, K.: Importance of lean management in a contemporary enterprise – research results. Res. Logist. Prod. **6**(3), 195–206 (2016)
5. Schwalbe, K.: Information Technology Project Management, 8th edn., Cengage Learning, US (2015)
6. Laudon, K.C., Laudon, J.P.: Management Information Systems: Managing the Digital Firm 15th edn., Pearson (2018)
7. Womack, J.P., Jones, D.T., Roos, D.: The Machine That Changed the World, Free Press, SKU: 9794 (1990)
8. Khmelevsky, Y., Li, X., Madnick, S.: Software development using agile and scrum in distributed teams. In: 2017 Annual IEEE International Systems Conference (SysCon) (2017)
9. Kheir andish, M.R.F., Khodashenas, H., Farkhondeh, K., Ebrahimi, F., Besharatifard, A.: An analysis on the evolution of Management Information Systems (MIS) and their new approaches. Interdisc. J. Contemp. Res. Bus. **4**(12), 491–495 (2013)
10. Betz, C.T.: Architecture and patterns for IT Service Management, Resource Planning, and Governance: Making Shoes for the Cobbler's Children, pp. 4–5. Morgan Kaufmann Publishers, Elsevier, San Francisco (2007)
11. Ignace, M.-P., Ignace, C., Médina, R., Contal, A.: la practique du Lean management dans l'IT, Pearson (2012)
12. Brand, K.: IT governance based on COBIT 4.1-A management guide, Van Haren Publishing (2008)

The Barriers to Lean Implementation in This New Industrial Context

Soukayna Belkadi$^{(\boxtimes)}$, Ilias Cherti, and Mohamed Bahaj

Department of Math and Computer Science,
Faculty of Sciences and Technology, Hassan 1st University,
Settat, Morocco
soukayna.belkadi@gmail.com

Abstract. This paper focuses on Lean's evolution, through a comparison between lean practices since its appearance at Toyota until today in new industrial context characterized by new scientific and technological challenges. We will specially focus on identifying the barriers that can hinder the application of lean in company, in this new industrial context framed by globalization, increased competition between support structures, technological progress, the emergence of information system and new customer that is becoming increasingly demanding.

We will therefore propose solutions that can help companies to overcome the obstacles that hinder the application of lean. Through five case studies that will allow us to collect a multi-stakeholder perception of barriers to lean.

Keywords: Lean thinking · Internal and external factors · Information system

1 Introduction

Since the success of the Japanese automaker Toyota, the Lean philosophy has inspired many companies, whether in industry or services. The goal of Lean is to optimize performance through continuous improvement and elimination of all sources of inefficiency. As a result, the new industrial context characterized by its rapid evolution, has prompted companies to move towards new improvement strategies. In fact, most organizations have more opportunities for improvement today than when they started their Lean philosophy ten years ago. The greatest opportunities for efficient implementation are those that still need to be discovered [2–5].

The originality of this work compared to the other research carried out is based on the combination of two major parts: the evolution of lean over the years and the obstacles to its implementation in a company, by relying on the opinions of several actors, from the top management to the operators, however a lot of previous work research were interested to the lean's obstacles but did not take into account the relevant evolution of lean thinking since its appearance until today, [1] and focused on the lean organizational and technical barriers and not the internal and external barriers to lean [4].

In this context this work contains two parts. In the first part, we define the concept of lean through an update on its evolution since its appearance at Toyota until today.

© Springer Nature Switzerland AG 2019
M. Ezziyani (Ed.): AI2SD 2018, AISC 913, pp. 214–221, 2019.
https://doi.org/10.1007/978-3-030-11881-5_18

The second part will be about the barriers to lean implementation though five case studies of five companies that operate in various industrial sectors: automotive, aeronautics, agro food industry, textile and cement plant. The last part presents the results of our research that we will discuss before proposing solutions for the obstacles that can hinder the evolution of lean concluding this article.

2 The Lean Between Today and Yesterday

The concept of "lean" dates back a long time and goes back to the 1950s, the origins of the lean thinking goes back to the Japanese manufacturers. (Shingo, 1981, Ohno, 1988, Monden, 1993). Much of this early work was carried out under the direction of Taiichi Ohno in the manufacture of car engines during the 1950s, later in the assembly of vehicles (1960s), and the wider supply chain (1970s) [6].

After 1990, the orientation of the workshop gradually widened, characterized by the design by many detractors. This evolution has focused on quality (early 1990s), cost and delivery (late 1990s), customer value from 2000, Toyota's philosophies have been shaped by the personalities, ethics and abilities of its creators in the Toyota family. Lean's principles are firmly grounded in scientific wisdom and methods (Bicheno and Holweg 2009) [3–6].

Currently, the customer orientation should be taken into account at all levels of the supply chain and needs to be implemented by all the staff of a company, from its managers to the line operators. To make a customer tied to a company's products, companies need to be increasingly flexible to be adapted to the new requirements of the customer that is becoming more and more demanding. We will outline the engineered improvement, scientific improvement, technology and globalization and industrial context that have impact the evolution of lean (see Fig. 1).

In recent decades, lean manufacturing has grown tremendously as a means of identifying waste, or activities and resources that do not add value to eliminate them. Consequently, the aim of the Lean concept is to increase the proportion of value added in the selling price of the product and thus reduce waste in all areas of the company.

Today, technological advances and external pressures are making technology less likely to go head-to-head with lean.

In the past, lean users were very satisfied with the results achieved through the adoption of simple lean tools in a stable industrial environment characterized by less competitiveness and complexity compared to today, which may help explain why some practitioners say that the most sophisticated tools that should be involved are paper, pencils and spreadsheets.

Today, the new industrial context framed by technological advances, the emergence of information system, external pressures from customers and prices make technology less likely to engage in face-to-face business today.

Lean tools have evolved over time to be adapted to this new industrial context, the use of technology and machine learning problems has become one of the parameters taken into account in lean thinking.

Currently, the application of a Lean management in a company can be associated with the development of information systems. But, conversely, the principle "think

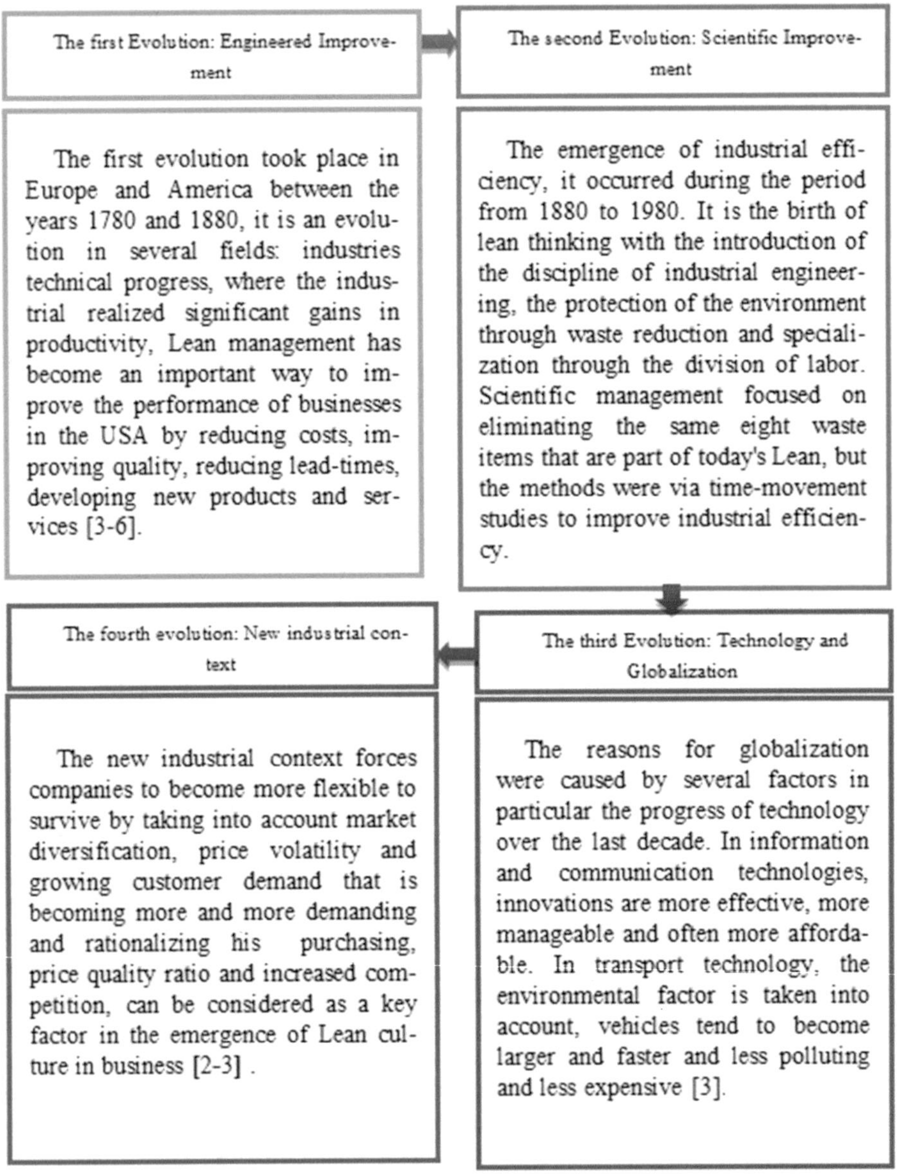

The first Evolution: Engineered Improvement	The second Evolution: Scientific Improvement
The first evolution took place in Europe and America between the years 1780 and 1880, it is an evolution in several fields: industries technical progress, where the industrial realized significant gains in productivity, Lean management has become an important way to improve the performance of businesses in the USA by reducing costs, improving quality, reducing lead-times, developing new products and services [3-6].	The emergence of industrial efficiency, it occurred during the period from 1880 to 1980. It is the birth of lean thinking with the introduction of the discipline of industrial engineering, the protection of the environment through waste reduction and specialization through the division of labor. Scientific management focused on eliminating the same eight waste items that are part of today's Lean, but the methods were via time-movement studies to improve industrial efficiency.

The fourth evolution: New industrial context	The third Evolution: Technology and Globalization
The new industrial context forces companies to become more flexible to survive by taking into account market diversification, price volatility and growing customer demand that is becoming more and more demanding and rationalizing his purchasing, price quality ratio and increased competition, can be considered as a key factor in the emergence of Lean culture in business [2-3] .	The reasons for globalization were caused by several factors in particular the progress of technology over the last decade. In information and communication technologies, innovations are more effective, more manageable and often more affordable. In transport technology, the environmental factor is taken into account, vehicles tend to become larger and faster and less polluting and less expensive [3].

Fig. 1. The evolution of lean throughout years

product, not project" is very disturbing for a team of developers who have not been trained and accustomed to this.

However, in order to define how to move from the methods used in the industrial world to those used in the information systems, the Information Systems Department should be reconsidered as a separate company: providing products and services to its customers with constraints management and administration (budget, human resources…).

While lean is an essential tool for the company for a better visibility of a production and to eliminate all activities without added value, its implementation can be slowed down by several obstacles, In the following, we will explain the obstacles that can hinder the establishment of lean culture within the company.

3   Case Study

We selected five leading companies from different industrial sectors that we thought were sufficiently representative. The five companies were selected according to their sector of activity and their application of the Lean approach. We conducted several interviews with people of different status who have experienced the implementation of Lean or who accompanied companies in this process, in order that our study will be dynamic we collected the strategic point of view of the top management of a company and the operational point of view expressed by the operators who lived the establishment of the Lean culture in their companies. All were realized with the help of an interview guide which is structured in many themes: the company and its organization, the strategic decision making concerning the adoption of Lean, the implementation phase of Lean and we're going to focus on the way that can help the company to overcome lean barriers.

Through the interviews and the questionnaire developed we have been able to mention several obstacles that can hinder the implementation of Lean within a company, these barriers are different from one position of responsibility to another, these obstacles will be developed in the following paragraph.

- Obstacles perceived in the setting up of the lean

Lean barriers can be caused by many factors, internal or external to the business. External barriers that include those related to the offer (lack of funding, obtaining technological information, raw materials, external environment, choice of suppliers, stakeholder-globalization), and to the request (consumer needs, their perception of the risk of innovation, limits of domestic and foreign markets) and the environment (government regulations, political actions). [1] Internal barriers are related to resources (technical expertise, management time, culture) and human nature (top management attitude, employee resistance to risk). They would correspond to all internal resources of the company, financial and human.

4   Results and Discussion

4.1   Internal Obstacles to Lean

Internal barriers related to human resources were those that were most raised in the questionnaire results and also in the interviews. Here again, companies that have successfully adopted Lean have a stronger and wider awareness of these endogenous barriers. The human resource obstacles are related to the non-comprehension of the Lean principles, which leads to resistance to change, thinking that Lean is a synonym for more and faster work in the execution of each task, especially the deletion of post by deleting non-creative tasks of added values. They are essentially related to the attitude towards change of employees as well as management or top management. This resistance to change comes mainly in the implementation phase as it can intervene more marginally in the phase of continued use. It would affect more widely employees who have seniority believing that their expertise is sufficient for effective work without resorting to Lean.

While we were analyzing the results of the questionnaire, one of the answers that attract our attention is that Lean is a tool for job cancellation, explaining that it succeeds in eliminating tasks and unnecessary operations. It represents a potential risk of job cancellation.

By analyzing his thinking It is important to keep in mind his poor understanding of the Lean principle and tools, hence the need for the intervention of the human resources department to schedule training sessions dedicated to all stakeholders who will be affected by Lean, before implementing Lean in a company, to explain that tasks and positions are two things strictly different, and especially that the purpose of Lean tools such as the value stream mapping used for analyzing the current state and designing a future state for the series of events that take a product or service from its beginning through to the customer by eliminating all flows without added value to allow employees to devote more time and energy to value-added tasks, in the security of their jobs.

"Lean brings a lot of standards that can hinder creativity" this answer explains that standards are often seen as constraints of creativity.

However creativity and innovation are one of the most important pillars of Lean, and among Lean practices is the Idea Fair, which encourages all employees of a company to bring innovative ideas through brainstorming. Lean standards are neither requirements nor laws, but rather baselines for measuring change. This change cannot be done without a certain level of creativity.

"When we talk about putting in place a tool or method to improve the quality of their work, most people are on the defensive. They think that these improvement initiatives will indicate that the quality of their work is insufficient. If this reaction is natural, it can however prevent an employee from developing his skills and progress. This misinterpretation of the purpose of Lean is one of the most difficult obstacles of the implementation of lean culture. However, analyzing and evaluating processes by using lean performance indicators is not intended to challenge the professionalism of employees. It is not a matter of controlling their competencies, but of giving each member of a company the opportunity to improve and identify where their energy is focused, through daily experiments and to encourage progress towards critical goals of the organization.

Otherwise, the misunderstanding of Lean management, push employees to think that Lean involves working ever faster, while speeding things up is not the goal of lean: Imagine that a process generates a lot of waste. Making it faster will just speed up the production of waste. In this example, Lean will seek to minimize this amount of waste rather than making the process faster. If a Lean approach usually results in accelerated processes, it is simply that the activities that were not needed have been removed by using a lot of tools such as SMED (Single-minute exchange of die) which provides a rapid and efficient way of converting a manufacturing process from running the current product to running the next product. This rapid changeover is key to reduce production lot sizes and thereby improving flow. So the lean concept aims on working smarter, not faster.

The question of time was also deployed in some answers to the questionnaire "Lean consumes a lot of time and effort for its implementation and generates more monitoring of the indicator, more monitoring of daily activity and therefore more support", "we are always in conflict with time to ensure that the product is delivered at the right time to customers, the implementation of Lean can help us certainly, but these results are not measurable immediately".

However the implementation of Lean cannot affect the productivity of employees, the finality is to remove the waste thus to save time by eliminating the useless tasks, for example using the Visual Factory Makes the state and condition of manufacturing processes easily accessible and very clear to everyone.

Furthermore the implementation of the various tools of Lean does not happen immediately and does not require a stop of chain of production; moreover the results of Lean cannot be seen directly after its implementation, such as such the VSM which provides a clear and simple vision of the process and the Total Productive Maintenance (TPM) which focuses on proactive and preventative maintenance to maximize the operational time of equipment. TPM blurs the distinction between maintenance and production by placing a strong emphasis on empowering operators to help maintain their equipment. Its principal goal is to create a shared responsibility for equipment that encourages greater involvement by plant floor workers. In the right environment this can be very effective in improving productivity (reducing cycle times, and eliminating defects).

The analysis of the case studies showed us that several internal obstacles can hinder the implementation of Lean in a company namely: non-adhesion and non-collaboration of the human resources, Underestimating employee attitudes, the non-motivation of the staff, the fear of responsibility (self-control for each operation), non-commitment of top management, lack of understanding Lean concept and resistance to change.

4.2 External Obstacles to Lean

The answers to the questionnaire did not talk about the external factors such as obtaining funding, partners, suppliers, globalization context etc. they do not see that these elements can hinder the implementation of Lean in the company, however external factors can also intervene in the implementation of Lean, as we already mentioned in the previous paragraph among the actions implemented by the companies to succeed the Lean implementation is the feedback of experience that allows to take advantage of the experiences of others companies that have already implemented Lean, it turns out that the external information flow from these companies can help to succeed the Lean experience and learn from the mistakes of other companies that have already done this exercise, looking for a partner also intervenes when company look for recruiting people having already recognized experience in Lean. The choice of suppliers can also impact the implementation of Lean, so it necessary to choose suppliers who already have Lean experience and who work with pull flow to align on the same just-in-time practices.

To overcome these problems some companies have set up a feedback system, to draw on the experience of companies that have already implemented the Lean culture in their company and to also benefit from the experience of employees who have previously worked with Lean in another company before joining their current position. These companies have chosen to involve all people who may be affected by Lean, especially the operators who have benefited from the training sessions scheduled by the Human Resources Department, these operators will participate in the definition of new working procedure based on Lean principle, this action will allow the company to overcome the obstacles related to resistance to change and Underestimating employee attitudes.

5   Conclusion

To sum up, The lean as we have just mentioned has evolved greatly between yesterday and today and has been able to be adapted more and more with the new industrial context, which requires a great deal of vigilance and flexibility on the part of companies to satisfy the needs of customer. Its implementation can be curbed by several obstacles that the company can overload by adopting a model that involves all their staff specially the top management that must motivate employees to get involved in the implementation of lean.

The Lean is all about "RESPECT FOR PEOPLE". Not just respect but love. Top management of companies need to love their employees enough (this relationship has to come from the top to the down) to cause real change. Change that allows companies to remove the waste so they can have a better work life balance.

As a perspective, our future work will seek to develop the external barriers to lean, in particular globalization, which has been accelerating since the mid-1980s and is impacting today a very large number of sectors as well as the preservation the environment which has become a key element of green lean taken into account throughout the supply chain.

Acknowledgement. At the end of this work, which is part of our thesis on lean thinking, we express our gratitude and our sincere thanks to all those who contributed directly or indirectly to the realization of this article in particular M. Cherti and M. Bahaj who assisted us during the various stages of our work. Their guidance and valuable advice and availability have allowed us to master and overcome the challenges of this communication.

References

1. Dubouloz, S.: Les barrières à l'innovation organisationnelle: Le cas du Lean Management. Manag. Int. **17**(4), 121–144 (2013). https://doi.org/10.7202/1020673ar
2. Grzelczak, A., Werner-Lewandowska, K.: Importance of lean management in a contemporary enterprise – research results. Poznan University of Technology **6**(3), 195–206 (2016)

3. Hines, P., Holweg, M., Rich, N.: Learning to evolve: A review of contemporary lean thinking. Int. J. Oper. Prod. Manag. **24**(10), 994–1011 (2004)
4. Lodgaarda, E., Ingvaldsena, J.A., Gammea, I., Aschehouga, S.: Barriers to lean implementation: perceptions of top managers, middle managers and workers. In: 49th CIRP Conference on Manufacturing Systems (CIRP-CMS 2016), Elsevier, vol. 57, pp. 595–600 (2016)
5. Ben Naylor, J., Naim, M.M., Berry, D.: Leagility: integrating the lean and agile manufacturing paradigms in the total supply chain. Int. J. Prod. Econ. **62**(1–2), 107–118 (1999)
6. Emiliani, M.L.: Origins of lean management in America: the role of Connecticut businesses. J. Manag. History **12**(2), 167–184 (2006)

Toward a Study of Environmental Impact of the Tangier Med Port Container Terminals

Fatima Ezzahra Sakhi[✉], Abdelmoula Ait Allal, Khalifa Mansouri, and Mohammed Qbadou

Laboratory: Signals, Distributed Systems and Artificial Intelligence (SSDIA)
ENSET Mohammedia, University Hassan II, Casablanca, Morocco
sakhi.fatimaezzahra@gmail.com,
aitallal.abdelmoula67@gmail.com,
khmansouri@hotmail.com, qbmedn7@gmail.com,

Abstract. The containerized cargo segment has continued its improvement during the last decades. This improvement has motivated the maritime stakeholders to invest in container terminal projects to boost their economies and to ensure the prosperity of their countries. However, the container terminal activity has a significant environmental impact on air, ocean, workers, and the surrounding population. The port of Tangier Med as one of the worldwide container terminal hubs does not make the exception in term of environmental impact. In this paper, the environmental impact of this port is quantified in term of greenhouse gases emission and pollutant particles. Innovative operational and technical solutions have been proposed to attenuate this impact and to make of it an environment friendly port.

Keywords: Environment · Port · Ship · Green energy · Rubber fragments · GHG emission

1 Introduction

The ocean transport remains the most important mode of transport for the worldwide economy trading. Over 80% of global trade by volume and more than 70% of its value being carried on board ships and handled by seaports worldwide. Consequently, the commercial shipping fleet has grown by 3,5 through the year 2016, reaching the number of 93161 vessels on 1 January 2017. In the worldwide economy trading, the containerized cargo is estimated at 245 609 000 dwt in 2017. This has been accompanied with an increase in number of deployed container ships, estimated at 11150 vessels and a world container port throughput increase by 1.9%, with volumes totaling 699.7 million TEUs. Whereas, in Africa the world container port has posted a throughput of 27 909 132 TEU, In 2016 [1]. This significant throughput increase has motivated the terminal operators and investors to reconsider their capacity expansion plans and the built of new container terminals.

Due to its advantageous geographical and commercial strategic position, Morocco does not make the exception. In the last decades, the Moroccan government did not stop investing in port infrastructures, to improve its worldwide trading position in term

© Springer Nature Switzerland AG 2019
M. Ezziyani (Ed.): AI2SD 2018, AISC 913, pp. 222–235, 2019.
https://doi.org/10.1007/978-3-030-11881-5_19

of logistic, competitiveness, and efficiency. Therefore, many port projects have emerged. One of the important projects is Tangier Med port, which is located on the Mediterranean Sea at 43.1 km east of Tangier city. Vu its strategic position, this port does not take a long time to become one of the hubs of the container shipping lines. Besides its benefit in term of economy development and prosperity of countries, the container ports have an environmental impact on air, oceans and coastal surrounding areas. This work consists of the assessment of this environmental impact. Operational and technical measures are proposed to attenuate this impact and make of this port one of the worldwide environment friendly container terminal hub.

The shipping environmental impact on oceans, air and land is one of the fertile research subjects that has drawn the attention of scientific and industrial researchers. In the available literatures, many researchers have been working on the subject in order to quantify this impact and propose adequate solutions to mitigate it. Winnes et al. [2] quantified the potential reductions in ship GHG emissions from port efforts, and analyzed ship emissions projections in the port area for 2030 in three scenarios, '1. Alternative fuel', '2. Ship design' and '3. Exploitation'. The analysis result showed that the GHG emissions from ships in the port are projected to increase by 40% to 2030 in a business as usual (BAU) scenario. The highest reductions were seen in the 'Operation' scenario where GHG emissions were 10% lower than the BAU level. Lindstad et al. [3] suggested associating damages and policies with ports, coastal areas possibly defined as Emission Control Areas (ECA) as in the North Sea and the Baltic, and open seas globally. He argued that it may be desirable to allow burning very dirty fuels at high seas, due to the cost advantages, the climate cooling benefits, and the limited ecosystem impacts. Zetterdahl et al. [4] showed by micro-composition analyzes that the soot particles emitted by the combustion of a low sulfur residual marine fuel oil (RMB30) often have no trace of sulfur compared to the particles from the combustion of heavy fuel oil (HFO), which always have a sulfur content greater than 1% m/m and it also showed that RMB30 reduced PM particulate emissions by up to 67%, SO_2 emissions by 80%, and decreased total volatile organic compound (VOC) emission. Bailey et al. [5], leaned towards that emphasizes the adoption of better technologies such as switching to cleaner versions of diesel fuel or completely the transition to alternative fuels such as natural gas or propane, and retrofitting, repowering, or retiring older diesel equipment and vehicles. Shore power for moored vessels, zero emission technologies such as fuel cells and automated container handling. Naciri et al. [6] shew that combining wind energy with a pumped hydroelectric storage system could be a vital solution to solve the electricity crisis in Morocco, when combined with a large-scale development with energy storage can lead to a significant reduction in energy production costs. However, our contribution consists of the study of environmental impact of the port of Tangier Med. The in-situ study in this port, shows that the containers operators use many handling equipment, which are powered by diesel engine, i.e. trucks, rubber-tired gantry cranes, and supporting service cars. The ship berthing assistance vessels i.e. tug boats and pilot boat use marine gasoil (MGO) for their propulsion. In addition, the study shows that the vessel berthing at Tangier Med continue to burn heavy fuel oil (HFO) for their needs in electrical energy and heating, in absence of any restriction in term of fuel Sulphur content limitation as applied in other ports. The study shows that, ships, handling equipment, tug boats, pilot boat produce a considerable amount of pollutant greenhouse gases (GHG), i.e. CO_2, SO_2, NOx, and pollutant particles. In this paper, these pollutants are quantified, based on the collected data and in-situ study. Innovative solutions have been proposed to ensure the environmental sustainability of the port and make of it a green port, key example in its region.

This paper is organized as follow: in Sect. 1, the paper is introduced by presenting the global shipping industry economic situation, focusing on containerized cargo segment and its environmental impact. In Sect. 2, the approach methodology and data collection are presented. In Sect. 3, an overview of the international maritime organization (IMO) and national regulations related to environmental pollution prevention are presented. In Sect. 4, the ocean, air and land pollutants, i.e. CO_2, SO_2, Nox, particles matter (PM) and acoustic noise which are produced by the container terminal activity are enumerated. In Sect. 5, the quantification of the environmental impact of the Tangier Med port container terminal pollutants is carried out. In Sect. 6, operational and technical solutions are proposed to mitigate or to attenuate this environmental impact. In Sect. 7, the paper is concluded.

2   Data Collection and Approach Methodology

For an accurate assessment of the environmental impact of container terminals, an in-situ study has been carried out including, port visits, ship visits, collection of supporting data from port partners. Where the onboard ships visits have been focused on the analysis of the relevant documentation, such as, voyage performance abstracts, Oil record books, bunker delivery notes, and the implemented environmental regulations. The type of fuels used in different ports of call has been assessed through the voyage performance abstracts. Green innovative solutions have been investigated to verify its suitability to be implemented in the port of Tangier Med.

3   National and International Environmental Regulations

For the last decades, the environment pollution prevention has become a key priority issue with a particular concern from the world community. Therefore, several regulations have been implemented to reduce the environmental impact of the maritime industry. Among these regulations, the international maritime organization (IMO) MARPOL convention for the prevention of pollution from Ships, i.e. Marpol Annex I, Marpol annex IV and Marpol annex V and Marpol annex VI. This Convention describe the restriction regarding the pollution by oil, garbage, sewage, and GHG emission. The shipowners must implement these regulations on board of their fleets. At the national level, the law n° 10-95 on the water, the decree n° 2-04-523 of January 24th, 2005 obliges the respect of the limit values fixed by joint orders of the governmental authorities. While noise pollution related legislation has recently been considered, such as within the European territorial waters where it is regulated by the implementation of two directives, i.e. the Water Framework Directive (2000) and the marine strategy framework directive (2010).

4   Container Terminals Environmental Pollutants

The container terminals contribute considerably in the increase of GHG emission and production of pollutant particles. This emission is mainly composed of CO2, SO2, NOx and pollutant particles, i.e. rubber fragments and particulate matters. In addition, the acoustic noise pollution, produced by the ships running equipment (Fig. 1).

Fig. 1. Tangier Med port environmental pollutants

4.1   Carbone Dioxide (CO2)

Carbon dioxide is a molecule composed of a carbon atom and two oxygen atoms, it is a colorless and odorless gas, which is weakly acidic and flammable. According to [7], the atmospheric concentration of carbon dioxide reached a peak in 2016 the temperature was 2 to 3 °C higher and the sea level was 10 to 20 m higher compared to the current level due to melting ice sheets. Jean-Claude et al. [8] stated that the increase in the concentration of CO2 in the atmosphere is the main responsible of the increase in global warming by the greenhouse effect and the ocean acidification. The CO2 concentration of atmospheric CO2 has risen from 280 ppm (pre-industrial value) at 388 ppm in 2010 and between 1906–2005 the average temperature on the surface of the Earth has increased by 0.74 °C. The CO2 has also, a major role in the acidification of the oceans, because the increase in the atmospheric CO2 rate leads to a decrease in the pH of the oceans. The engine CO2 production is calculated by using the Eq. (1) [9].

$$M_{CO2} = D_C P_{Ci} E_f F_0 \tag{1}$$

Where,

M_{CO2}: is the mass of produced CO2.

Dc: Fuel flow in [t] or the annual consumption.

P_{ci}: Lower Calorific value in [TJ/t]: the amount of heat released by the complete combustion of a fuel unit determined by density and sulfur content.

E_f: Emission factor in [tCO2/TJ] this factor is used to transform a physical activity data into quantity of greenhouse gas emissions.

F_o: Oxidation factor of the fuel.

4.2   Sulfur Dioxide (SO2)

Sulfur dioxide is a colorless gas with a pungent odor, consisting of two oxygen atoms and one sulfur atom. The main anthropogenic source of SO2 is the combustion of fossil fuels containing sulfur. This gaseous pollutant acidifier contributes to the acidification of the environment once emitted into the air and in the presence of water it turns into sulfuric acid H2SO4 which contributes to the phenomenon of acid rain that can lead to the death of fish, the deterioration of buildings and soils and may have adverse effects on vegetation and plants. The Group of Experts on the Scientific Aspects of Marine Pollution (GESAMP) [10] indicated in its report that the emission of various pollutants such as SO_2, NOx and particles increase the acidification of the ocean. Because the absorption of gases such as sulfur and nitric acids leads to the decrease of the pH of the seawater, which in turn causes the release of CO_2 from the ocean to the atmosphere. Therefore, the more acidic the seawater is, less natural CO_2 it provides. SO_2 also has negative health impacts, short exposures to high values (more than 250 µg/m^3) can cause bronchial spasm, coughing, impaired breathing function and eye irritation [11]. The produced amount of SO_2 is calculated by using the Eq. (2) [12].

$$M_{SO2} = D_C P_{Ci} E_f \tag{2}$$

Where,

M_{SO2}: is the mass of produced SO2.

Dc: Fuel flow in [t] or the annual consumption.

P_{ci}: Lower Calorific value in [TJ/t]: the amount of heat released by the complete combustion of a fuel unit determined by density and sulfur content.

E_f: Emission factor in [tSO2/TJ] this factor is used to transform a physical activity data into quantity of greenhouse gas emissions.

4.3   Oxides of Nitrogen (NOx)

NOx Nitrogen Oxide is an irritant gas, contains two molecules nitrogen monoxide NO and nitrogen dioxide NO_2, it comes mainly from processes operating at high temperature such as combustion plants and industrial processes, it has adverse effects on health and the environment. Among these major environmental effects, acidification of the environment, which can lead to leaf or needle falls, necrosis and significantly influence aquatic environments and eutrophication, which is an excess intake of nitrogen in natural environments, particularly soils, leading in the reduction if biodiversity, under the effect of solar radiation, NOx associates with volatile organic compounds (VOCs) generates the formation of ozone in the lower layers of the atmosphere "tropospheric ozone". On health, NOx cause breathing difficulties or airway hyper-responsiveness in sensitive people and promote increased sensitivity of the bronchi to infections in children because it penetrates the finest branches of the airways which can lead to premature mortality and asthma attacks [13]. The produced amount of NOx is calculated by using the Eq. (3) [12].

$$M_{NOx} = D_C P_{Ci} E_f \qquad (3)$$

Where,

M_{NOx}: is the mass of produced NOx.
Dc: Fuel flow in [t] or the annual consumption.
P_{ci}: Lower Calorific value in [TJ/t]: the amount of heat released by the complete combustion of a fuel unit determined by density and sulfur content.
E_f: Emission factor in [tNOx/TJ] this factor is used to transform a physical activity data into quantity of greenhouse gas emissions.

4.4 Pollutant Particles

The presence of pollutant particles in the container terminal is due to the ship and port equipment engines exhausts, i.e. particulate matters (PM) and tire fragment. There are two types of PM: PM 2.5 are particles with a diameter of less than 2.5 μ and PM10 with a diameter of less than 10 μ. These particles are dangerous for health because they can easily penetrate deep within our breathing voices. The fine PM10 particles generated by the tires contain high concentrations of Zn, Cu, sulfur, PAHs (polyaromatic hydrocarbons), benzothiazoles, natural resins, and n-alkanes (saturated hydrocarbons) which have impacts on air quality and health. Whereas the rubber fragment is produced by the fair wear and tear of the rubber tires. This wear produces micrometric particles, which are harmful to the marine and terrestrial environment, activators. However, the leachate toxicity of these particles is due to zinc and organic compounds [14]. In [15] state that ambient road air contains approximately 0.4 to 11 mg/m^3 of tire particles and can be transported over relatively long distances. Large, which can cause terrestrial toxicity because the insertion of zinc into the soil generates an increase in its pH and therefore its toxicity which slowed the growth of plants (Smolders and Degryse, 2002). They also have negative effects on men's health, because exposure of human lung epithelial cells (A549) to organic extracts of tire particles has caused an increase in cell death and damage to DNA, as well as a significant change in cell morphology (Gualtieri et al., 2005a). The water from the first runoff transports the tire wear particles to the coastal waters and generates the toxicity of the latter which threatens the aquatic organisms, because the rubber leaching compounds have mutagenic, teratogenic and estrogenic effects on these species, and they are an important source of phthalates for rainwater.

4.5 Acoustic Noise

During berthing, the ships keep run their noisy engines, ventilation, and other auxiliary equipment. These engines generate a huge noise that spread easily in the water because it has a low resistance to the transmission of sound. According to Thirion et al. [16], the noise generated by auxiliary engines is generally characterized by a frequency below 1000 Hz, these frequencies are located in the audible scale of the majority of fish which decreases their hearing abilities and increases their stresses. In fact, about 700 aquatic species communicate by sound, and the majority uses it to orientate themselves, locate a partner, find their food and avoid predators. However, ship noise collides with the natural sound of the ocean, disrupts the acoustic environment, and impedes communication between these animals. The noise pollution can even lead to the death of some species if it exceeds 180 dB, because the intensity of the sound may cause the injury and the loss of the animal hearing, and the noise violence shock may lead to the immediate death by internal bleeding [16].

5 Tangier Med Port Container Terminals Environmental Impact Quantification

5.1 Tangier Med Port Overview

In 2017, Tangier Med port activity has evolved by 15% compared to 2016, with a volume of 51,328,150 tons of processed goods and a value of 88 billion MAD of Moroccan exported products. This trend is reflected in a 3% increase in passenger and ro-ro traffic, a 1% increase in the hydrocarbon traffic, a 16% increase in vehicle traffic at the Renault dedicated terminal and a 17% increase in container ships traffic with the berthing of 13,502 ships including mega-ship stopovers and passenger ships and RO-RO [17]. This change is also due to the changes and trends that affect this port, such as strengthening port infrastructure, opening up new business partners and developing regular lines. To improve the port performance, others areas need to be reconsidered in respect of safety, security, and environmental issues to ensure its competitiveness in a fluctuating shipping market. In fact, the port Tangier Med adopts an environmental approach for the preservation and protection of the environment which is based on the protection and preservation of biodiversity, energy efficiency, eco-responsible management and renewable energies. Despite these achievements, the port still has failures in this environmental segment, as it is the case for the majority of the worldwide container terminals. The container terminals serve approximately 3911 container ships par year. The stay of these ships in the port lead to the production of GHG, acoustic noise and particulate matters (PM) which are thrown from the ships exhaust funnels. In addition to pollutants which are produced by the ships, the cargo handling equipment produce also GHG, PM and tires' rubber fragments (Fig. 2).

OD: Overall diameter
OW: Overall width
OTD: Original tread depth

Fig. 2. Tire dimensions designation

5.2 Container Terminals Cargo Handling Equipment GHG Production

The activity of the port is supported by various equipment, used for cargo handling and ships berthing assistance. This equipment is powered by diesel engine that burn gasoil. The Table 1 presents the terminal 1 and 2 cargo handling equipment and its production in term of GHG emission. However, the container gantry cranes are excluded from the study because they are powered by electrical energy. The container terminal environment impact assessment shows that the total annual consumption of Gaz oil is of the order of 8570 MT, producing approximately 26368 MT of CO2, 16.1 MT of SO2 and 514 MT of NOx.

5.3 Container Terminals Cargo Handling Equipment Rubber Fragments Production

The cargo handling equipment such as, rubber tyred gantries (RTG), reach stackers, terminal trucks and empty handler are rubber tired. The tires fair wear and tear results in the production of considerable amount of rubber fragments. This amount is spread over the cargo handling area (Fig. 3) where the rain may take it to the sea or may be moved to other surrounding areas. The amount produced depends on the daily use of the equipment and the quality of the tires.

Table 1. Port facilities GHG emission quantification

Equipment	Terminal 1	Terminal 2	Annual Gasoil consumption (MT)	Annual production of CO2 (MT)	Annual production of SO2 (MT)	Annual production of NOx (MT)
Rubber Tyred Gantries	23	21	6170	18984	11.6	370
Terminal Trucks	48	36				
Reach Stackers	2	2				
Empty Handlers	5	2				
Forklifts	3	3				
Emergency generator	2					
Tugboat	4		2400	7384	4.5	144
Pilot boat	2					
Total			8570	26368	16.1	514

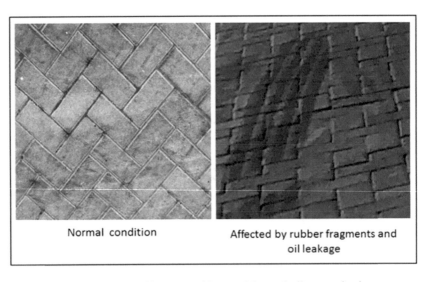

Normal condition   Affected by rubber fragments and oil leakage

Fig. 3. Cargo handling area rubber particles and oil contamination

The assessment result is presented in Table 2 where is shown that the container terminals produce a 15 m^3 and 23400 kg of rubber fragments. As the specific gravity of rubber is arrange from 920–2200 kg/m^3. An average of 1560 kg/m^3 is used for this calculus. The rubber fragments production is calculated by using the developed Eq. (4).

$$M = \pi \rho O_w O_{dt} N_t (O_d - O_{dt}) \tag{4}$$

Where,

M: mass of produced rubber fragments in (Kg),
ρ: rubber specific gravity in (Kg/m3),
O_d: tire overall diameter in (m),
O_w: tire overall width (m),
O_{dt}: original tread depth (m),
N_t: Number of worn tire per year.

Table 2. Cargo handling equipment rubber fragment production

Equipment	Tire size	Tire overall diameter (mm)	Tire overall width (mm)	Original tread depth (mm)	Annual tire consumption	Annual Worn rubber volume (m^3)	Annual worn rubber weight (Kg)
Rubber Tyred Gantries	18.00–25	1750	570	51	24.00	3.78	5896.80
Reach Stackers	18.00–25	1750	570	51	27.00	4.25	6630.00
Terminal Trucks	295/80R22.5	1048	299	16	960.00	15.00	23400.00
Empty Handlers	14.00–24	1350	373	25.5	24.00	0.96	1497.60
Total						23.99	37424.40

5.4 Tangier Med Port Container Terminals Calling Ships GHG Production

When the port is congested, the ships has to drift or to drop the anchor keeping the main engine running at idle speed to be ready for maneuvering, and when the ship enters or leaves the port, the main engine is kept running at idle speed to respond to request maneuvering. Once the ship is alongside, the main engine is stopped. Whereas, the auxiliary engines and boiler are kept running to ensure electrical energy production and heating steam production. In both situations, the ship equipment consumes heavy fuel oil for its powering. This lead to production GHG, and production of particulate matters. The study shows that the ship stays alongside for an average period of 0.56 day, with an HFO consumption of 6.00 MT/day. While, the ship at idle, stays outside the port for an average period of 1.45 day, with an HFO consumption of 10.5 MT/day. The container terminals ships GHG emission assessment result is presented in Table 3, where it is shown that the total annual consumption of HFO is of the order of 72871.73 MT, resulting in the production of 262461.74 MT of CO_2, 5293.35 MT of SO_2 and 4741.95 MT of NOx.

Table 3. Calling ships GHG emission quantification

Number of container ship	Annual Average stay (day)	Annual average idle (day)	Annual stay HFO consumption (MT)	Annual idle HFO consumption (MT)	Total HFO consumption	CO2 production (MT)	SO2 production (MT)	NOx production (MT)
3911	2189.98	5670.95	13326.73	59545.97	72871.73	262461.74	5293.35	4741.95

6 Innovative Operational and Technical Measure to Attenuate the Port Environmental Impact

6.1 Use of Green Energy to Attenuate the Port Environmental Impact

Tangier Med port has many environmental credentials to be a green port. Among these qualifications, is its proximity to the dam Oued Rmel and the availability of wind energy. The Fig. 4, shows the possibility to supply the port facilities and ships with green energy. It consists of pumped hydro-electrical storage system. This system permits the storage of wind energy and to restitute in demand. In the energy storage phase, the hydroelectric machine operates in pump-motor mode, it consumes electricity generated by the wind station to pump water from the lower basin to the upper basin and store wind energy in the form of gravitational potential energy. While in restitution phase, the hydroelectric machine operates in a turbine-generator mode and converts the gravitational potential energy of the water into electricity during the transfer of water from the upper basin to the lower basin.

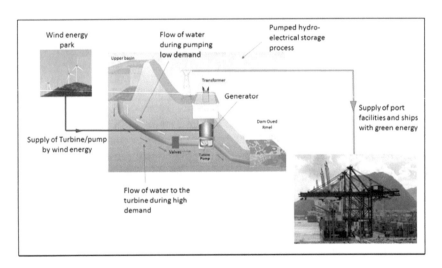

Fig. 4. Tangier Med port green energy supply concept

6.2 Use of Electrical Driven Equipment

The use of electric motorized means of handling instead of those which operate by the combustion of fossil energy is a good alternative for an ecological terrestrial transport within the Tangier Med port especially if this electrical energy is produced by renewable energies. According to Yang et al. [18], the conversion of the handling means exiting with a diesel oil to the electrical energy allows energy savings of 86, 60%, 67,79% reduction of CO2 emissions, reduction of noise pollution and reduction of maintenance costs and downtime. Electric vehicles remain a transmitter of fine particles produced by tire abrasion due to rolling, road friction and braking. The use of good quality "green tire" tires in these areas is essential, as a quality tire can last twice as long as a poor-quality tire, which allows a low rolling resistance and therefore a fuel-saving. According to Michelin green tires despite their cost, they allow a saving of 0.2 l of fuel/100 km, 4 g of CO2, and 125 euros over the average life of tires.

6.3 Implementation of Autonomous Ship as Shipping Alternatives

An autonomous vessel is a vessel that operates autonomously or remotely controlled from shore control center, without crew on board. This is a concept that brings together autonomy, remote control, and automation. As the autonomous ship, will operate without crew onboard, the accommodation and crew life being facilities are no longer needed. The elimination of crew onboard will reduce considerably the energy consumption, production of garbage and sewage. The "non-ballast design" will eliminate the ballasting system related consumers (pumps, control systems) [19]. It is assumed that the AS will use clean energy for its propulsion such as liquefied natural gas (LNG), distilled FO, GO, and electrical propulsion. The benchmark of autonomous ship and conventional ship in term of energy saving shows a saving of 74.5% of the energy consumed by adoption of new design and the elimination the listed equipment [20]. The classification society DNVGL autonomous ship concept called ReVolt is powered by 3000 Kwh battery, resulting in reduction of operating cost by minimizing the rotational components maintenance cost. The vessel has a range of 100 nautical miles battery charge autonomy and compared to diesel propulsion, might save up to USD 34 million [21]. The implementation of this type of ship in the port of Tangier Med will result in energy-saving and GHG emission reduction.

7 Conclusion

The worldwide development in containerized cargo has boosted the expand of container terminals all over the world. Beside its obvious benefits in economy improvement and countries prosperity, this container terminal expanding has a considerable environmental impact on air, ocean and human. In this paper, the assessment and quantification of the environmental impact of Tangier Med port container terminals show that this port contributes significantly in the production of GHG, i.e. CO2, SO2, NOx, MP, and production of rubber fragments. To mitigate or attenuate this impact, innovative green operational and technical measures have been proposed. These measures consist of providing electrical cargo handling equipment, electrical propelled tugboat, electrical propelled pilot boat and to supply them with wind energy. The other measure is to supply the ships which are alongside with electrical shore connection to allow stoppage of auxiliary engine. These innovative solutions permit to the port to be converted in green port and to make of it a key example of pollution prevention in its region. The perspective of this work is the study of the implementation of autonomous ship in maritime supply chain to ensure the maritime environmental sustainability.

References

1. United Nations Conference on Trade and Development (UNCTAD); Review of maritime transport 2017; UNCTAD/RMT/2017, United Nations Publication (2017). ISSN 0566-7682
2. Winnes, H., Styhre, L., Fridell, E.: Reducing GHG emissions from ships in port areas. J. Res. Transp. Bus. Manage. **17**, 73–82 (2015)
3. Lindstad, H., Eskeland, G.S., Psaraftis, H., Sandaas, I., Strømman, A.H.: Maritime shipping and emissions: a three-layered, damage-based approach. Ocean Eng. **110**, 94–101 (2015)
4. Zetterdahl, M., Moldanova, J., Pei, X., Pathak, R.K., Demirdjian, B.: Impact of the 0.1% fuel sulfur content limit in SECA on particle and gaseous emissions from marine vessels. J. Atmos. Environ. **145**, 338–345 (2016)
5. Bailey, D., Solomon, G.: Pollution prevention at ports: clearing the air. Environ. Impact Assess. Rev. **24**, 749–774 (2004)
6. Naciri, M., Aggour, M., Ait Ahmed, W.: Wind energy storage by pumped hydro station. J. Energy Syst. **1**, 32 (2017)
7. Organisation météorologique mondiale, OMM-N° 1167 (2016)
8. Jean-Claude, K., Pierre-André, H.: The consequences of an increase of the atmospheric CO2 concentration, a global survey of our current state of knowledge, World engineer's convention (2011)
9. La Ministre de l'ecologie et du developpment durable (France). Official Journal of 6 December 2005, NOR:DEVP0540388A. http://www.bulletin-officiel.developpement-durable.gouv.fr/fiches/exboenvireco/200601/A0010017.htm
10. Group of Experts on the Scientific Aspects of Marine Pollution (GESAMP). The atmospheric input of chemicals to the ocean, GSAMP reports and studies No. 84 (2012)
11. Gouvernement du Grand-Duche de Luxembourg, Effects of SO2 on health and the environment (2017). www.environnement.public.lu/fr/loft/air/Polluants_atmospheriques/les_oxydes_de_soufre_SOx/effets-SO2.html
12. Fédération francaise de l'acier, Guide méthodologique pour l'évaluation des émissions dans l'air des installations de production et de transformation de l'acier, Version 5, Décembre 2004
13. Buchet, E., Dubeau, B., Guerin, A.: Study of the energy impact and emissions of the ports and airports of the PACA region (2012)
14. Turner, A., Rice, L.: Toxicity of tire wear particle leachate to the marine macroalga, Ulva Lactuca. J. Environ. Pollut. **158**, 3650–3654 (2010)
15. Wik, A., Dave, G.: Occurrence and effects of tire wear particles in the environment – a critical review and an initial risk assessment. J. Environ. Pollut. **157**, 1–11 (2009)
16. Thirion, J.M., Doré, F., Sériot, J.: Impact de la pollution sonore sur la faune. J. Le courrier de la nature **254**, 32–37 (2010)
17. Tanger Med port complex, Activity – 2017 activity report (2017)

18. Yang, Y.-C., Chang, W.-M.: Impacts of electric rubber-tired gantries on green port performance (2013)
19. Ait Allal, A., Mansouri, K., Youssfi, M., Qbadou, M.: Toward a study of ocean pollution by maritime industry and proposal of innovative ideas to reduce its impact. In: Virtual Conference on Advanced Research in Materials and Environmental Science, 10 January 2018
20. Ait Allal, A., Mansouri, K., Youssfi, M., Qbadou, M.: Toward energy saving and environmental protection by implementation of autonomous ship. In: 19th IEEE Mediterranean Electronical Conference, IEEE MELECON 2018, 2nd–4th May 2018, Marrakech Morocco (2018)
21. DNVGL. The ReVolt new inspirational concept (2017). https://www.dnvgl.com/technology-innovation/revolt/index.html

A Copulas Approach for Forecasting the Rainfall

Adelhak Zoglat[1(✉)], Amine Amar[2], Fadoua Badaoui[3], and Laila Ait Hassou[1]

[1] Laboratory of Mathematics, Statistics and Applications, Faculty of Sciences,
Mohammed V University in Rabat, Rabat, Morocco
azoglat@gmail.com, laithassou@gmail.com
[2] Moroccan Agency for Sustainable Energy, Rabat, Morocco
[3] Department of Statistics, Demography and Actuarial Sciences,
National Institute of Statistics and Applied Economics Rabat, Rabat, Morocco

Abstract. Rainfall forecasting is a crucial issue in a semi-arid country like Morocco. Information on rainfalls can be used by marketers in the short term, to plan customer allocations and storage requirements. In the middle term, it can provide guidelines for seasonal selection of crops. In the long term, rainfall forecasts are important for hydrologists and water managers to build integrated strategies against potential disasters caused by tremendous floods or severe droughts.

Rainfall forecasting methods are of two kinds. The first one relies on statistical approaches while the second one is based on numerical simulations. Despite their higher cost, numerical simulations are still unable to consistently outperform simple statistical prediction systems. This is essentially related to the uncertainty of the relationship between rainfalls, hydro-climatic variables and climatic variability indices.

This paper aims to present a statistical approach supporting the use of lagged Southern Oscillation Index (SOI) for forecasting seasonal rainfall. We establish a statistical model, based on copulas theory, which takes the SOI and the rainfall relationship into account. Data are obtained from the World Meteorological Organization (WMO), for 6 meteorological stations located in Morocco (Tangier and Casablanca), Spain (La Coruna and Valladolid), and Portugal (Lisboa-Geofísica and Santa Maria). Using the suggested approach, we conduct a deep temporal and regional comparative analysis, which leads to the adjustment of different families of copulas, but with a predominance of Normal and Clayton copulas. For all stations and seasons, final results confirm a delayed effect in the structure between Rainfall and SOI with a strong relationship on the central part, while rainfall extreme events can be related to other atmospheric and climatologic indices.

Keywords: Rainfall forecasting · Southern Oscillation Index (SOI) ·
Copulas theory · Quantile regression

© Springer Nature Switzerland AG 2019
M. Ezziyyani (Ed.): AI2SD 2018, AISC 913, pp. 236–244, 2019.
https://doi.org/10.1007/978-3-030-11881-5_20

1   Introduction

Forecasting rainfalls constitutes a highly interesting issue, due to interferences with many economic and social fields and also to the multiplication of climate change impacts. This issue is first a matter of a continuous observing process which concerns the state of the atmosphere, oceans and land surface. In this context, the World Meteorological Organization (WMO) provides a framework for an advancing suite of worldwide observing systems based on satellites, radars, and surface weather stations. Scientists have made considerable progress in developing statistical and mathematical techniques to integrate these observations for weather prediction on scales from individual clouds to regional severe weather events and global patterns.

For the "medium and long" term (a few hours to a month or more) forecasts, numerical weather prediction (NWP) is presented as the dominant forecasting technique. The NWP involves the representation of the current atmospherics state on a three-dimensional grid, applying the physical and dynamical equations that govern how the atmosphere changes in time at each grid point. Repeating this process generates a forecast of desired length. Despite their sophistication, NWP models suffer from short comings, and insufficient computing resources. Imperfect simulation of small-scale phenomena, such as clouds and precipitation, due to the chaotic nature of the atmosphere and errors in the modeling system is another inconvenient of NWP (American Meteorological Society, 2015 [1]).

Many studies are currently interested in statistical approaches, and use lagged oceanic and atmospheric climate indices to accurately calculate probabilities of receiving particular amounts of rainfalls, at a particular location and over next few months. In this context, the Southern Oscillation, which describes changes in temperature of the ocean waters and air pressure in the Pacific Ocean, is presented as a potential indicator when forecasting rainfalls. The Southern Oscillation can be defined more precisely as the see-saw pattern of reversing surface air pressure between the eastern and western tropical Pacific Ocean. So, when the surface pressure is high in the eastern tropical Pacific Ocean, it is low in the western tropical Pacific Ocean and vice versa. The Southern Oscillation Index (SOI) is calculated from the monthly or seasonal fluctuations in the air pressure difference of the area between Tahiti (in the mid-Pacific) and Darwin (in Australia). It gives a simple measure of the strength and phase of the difference in sea-level pressure, given in terms of an index. Thus, a strong and consistent negative SOI pattern is related to El Niño and conversely, a deep and consistent positive SOI pattern is usually associated with below normal rainfall and La Niña.

Because of its key role in different socioeconomic sectors, rainfalls forecasting is a very active research area. The literature abounds with papers presenting different approaches to study problems related to rainfall. Chawla et al. [2] propose an assessment of the Weather Research and Forecasting (WRF) model for simulation of extreme rainfall events in the upper Ganga Basin. Pham et al. [3] have developed a coupled stochastic rainfall-evapotranspiration model for hydrological impact analysis. Jha et al. [4] use a Bayesian post-processing approach to improve precipitation forecasts in a Canadian catchment. Zhang et al. [5] propose a Monte Carlo approach to forecast landslide disasters associated with heavy rainfalls. Kim et al. [6] use a copula-based Bayesian network to forecast quarterly inflow to reservoirs of Soyanggang and Andong dams in the republic of Korea. Sene et al. [7] use stochastic dynamic regression and transfer function approaches to forecast potential seasonal flow for large lakes.

The choice of the lagged SOI is discussed in many current studies on weather forecasting. It is used to predict Australian seasonal rainfall in some regions and seasons (Chiew et al. [8], McBride and Nicholls [9], Stone et al. [10], Kirono et al. [11]), to forecast early summer rainfall in the Lesotho Low lands (Hyden and Sekoli [12]), the mean annual rainfall in the North of Iran (Hadiani et al. [13]), the annual rainfall and drought in Zimbabwe (Chifurira and Chikobvu [14]), and to test a crop model (APSIM) with a skillful seasonal climate forecasting system to inform crop designs (Rodriguez et al. [15]). Zahmatkesh et al. [16] proposed to forecast real-time rainfall using different number of spatial inputs with different orders of lags. Agilan and Umamahesh [17] use El Niño Southern Oscillation cycle indicator for modeling extreme rainfall intensity over India.

In the same context, this paper aims to present a statistical approach supporting the use of lagged SOI for forecasting seasonal rainfall. Using a copula approach, we establish a statistical model that takes the SOI and the rainfall dependence into account. This model is used to conduct a comparative analysis using a data set obtained from the WMO.

2   Data, Methodology and Results

To explore the relationship between the lagged SOI and rainfall, we use data from 6 meteorological stations (Fig. 1) located in Morocco (Tangier and Casablanca), Spain (La Coruña and Valladolid), and Portugal (Lisboa-Geofísica and Santa Maria). The choice of the Mediterranean region is justified by the complexity of its orography and land-sea distribution, which explains the presence of homogeneous and persistent inter-seasonal variability in patterns. Moreover, the idea behind choosing six different locations aims at confirming or disconfirming the hypothesis of delayed effects according to locations, and to identify the most relevant months in the analyzed relationship. Our approach can be summarized in the following algorithm steps:

Step 1: Gather SOI and rainfall monthly series (SOI_t, $Rainfall_t$) and calculate, for each month t, the Kendall correlation coefficient between SOI_t and $Rainfall_t$. The Kendall correlation coefficient presents the advantage to capture general forms of dependence. At the end of this step, we retain only the "strongly correlated" series.

Step 2: Use a semi-parametric approach to identify families of copulas susceptible to fit the data. This consists in choosing the "closest" copula to Deheuvels empirical copula [18] using the metric based on the Mean Squared Error (MSE). The implementation of this approach is as follows:

1. Construct the empirical copula (for more details on this step, see Roncalli et al. [19]).
2. Use the scatter plot of Rainfall$_t$ against SOI$_t$ to have a preliminary idea on susceptible fitting copulas.
3. Among these susceptible fitting copulas, select the "closest" (for the MSE metric) copula to Deheuvels empirical copula.

To confirm the semi-parametric approach choice, we perform some parametric methods. The selection methods used in Step1 and Step 2 are quite subjective. To evaluate our choices adequacy we use some parametric methods (see Embrechts et al. [20] for more details).

Step 3: Test the adjustment quality of susceptible fitting copulas, using some goodness-of-fit tests.

A deep study of the correlations confirms the hypothesis of a delayed effect of SOI on rainfalls. For instance, SOI_{August} has a significant effect on $Rainfall_{October}$ in Tangier and Valladolid stations while $Rainfall_{October}$ in Casablanca and La Coruña stations are impacted by $SOI_{September}$ and SOI_{June}, respectively. However, it should be noted that SOI_{March}, has an immediate impact on rainfall in Tangier, Casablanca, Valladolid and La Coruña stations. No significant correlation has been detected for Santa Maria station. This is probably due to geographic characteristics. Santa Maria is an island forming the eastern group of the archipelago of the Azores with the island of São Miguel in Portugal. The scatter plots in Fig. 2 highlight two different patterns. For some series (the first row of Fig. 2), we can notice a positive correlation between the SOI and rainfall series. In contrast, for other months (the second row in Fig. 2), the SOI and rainfall series are negatively correlated.

Table 1 summarizes our selections of the best fitting copulas using the Anderson Darling test. The analysis of the daily bivariate series (SOI, Rainfall) for each month leads to different families of copula candidates. However, we can distinguish some form of stability. In fact, for all the considered stations, Normal copulas are the most appropriate to capture the SOI and rainfall relationship during March. For SOI_{August} and $SOI_{September}$, independently of the month and the station of rainfall series, Clayton copula is the most suitable. Note that a Normal copula belongs to Elliptical copulas family and thus doesn't exhibit extreme events. While the Clayton copula, a member of Archimedean copulas family, exhibits lower tail dependence. This can be thought of as a strong probability of extreme low rainfall during October and extreme low value of SOI during August and September.

These results show that the correlation is periods sensitive, and second the dependence structure between Rainfall and SOI is strong on the central part. Rainfall extreme events are probably related to other atmospheric and climatologic indices.

To test the stability of fitted copulas, we use them for forecasting seasonal rainfall. The results are reproduced in Fig. 3. It shows, for each station, the plots

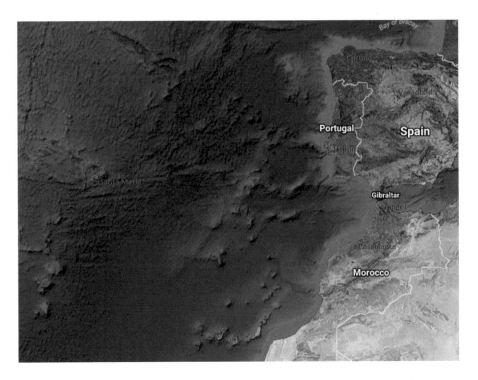

Fig. 1. Locations of the stations in our study (from google earth)

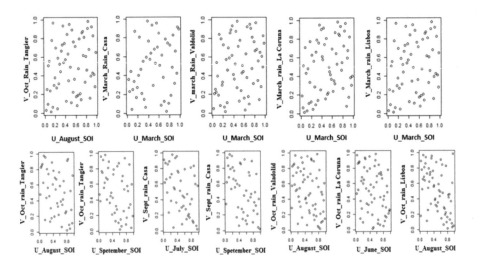

Fig. 2. Representation of joint marginal distribution for the retained months

Table 1. The final retained copulas

Station	Series	Fitting copulas	$P - value$
Tangier	$SOI_{March}; Rain_{March}$	Normal	0.7318
	$SOI_{August}; Rain_{October}$	Clayton	0.6800
	$SOI_{September}; Rain_{October}$	Clayton	0.8497
Casa	$SOI_{Jully}; Rain_{September}$	Frank	0.4830
	$SOI_{September}; Rain_{October}$	Clayton	0.6898
	$SOI_{March}; Rain_{March}$	Gumbel	0.3172
Valladolid	$SOI_{March}; Rain_{March}$	Normal	0.6019
	$SOI_{August}; Rain_{October}$	Clayton	0.8836
La Coru/rm̃na	$SOI_{March}; Rain_{March}$	Normal	0.8506
	$SOI_{June}; Rain_{October}$	Normal	0.8706
Lisboa	$SOI_{March}; Rain_{March}$	Normal	0.7633
	$SOI_{August}; Rain_{October}$	Clayton	0.8207

of the observed values (in blue) and the values predicted using a copula quantile regression (in green). For instance, the first plot represents the rainfalls in Tangier's station during March 2001, 2003, and 2004 ($Rainfall_{March}, Tangier$) together with the corresponding values predicted using a Normal copula quantile regression (Normal_ med). The term "med" signifies the median level used in the copula quantile regression approach.

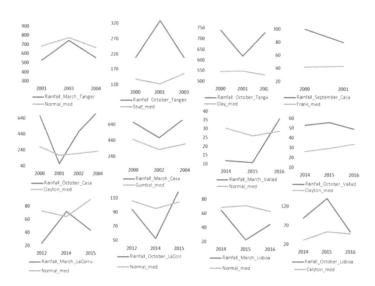

Fig. 3. Performances of the adjusted copulas in terms of the rainfall forecasting, using a copula quantile regression approach

Forecasted results are much contrasted. For some cases, such as rainfall during October in Valladolid and during March in Lisboa. The fitted copulas can't even predict the trend toward more or less rainfalls and then, the predicted values are not consistent with observed ones. In fact, the analysis of the observed values in some cases, confirms an upward (or downward) trend in the amount of rainfall while the predicted values show an opposite trend. In addition, the forecasting errors measured as the difference between observed and predicted values, can exceed for some cases 60% of the observed amount of rainfall. To handle such a situation a multivariate copulas approach involving more than two series is probably more appropriate. Indeed, that rainfalls are correlated with several atmospheric, geographic and environmental conditions.

3   Conclusion

Rainfall is a very important climate parameter that affects social activities and influences the performance of the economy over the world. It is closely associated with droughts and floods, food insecurity, energy and water shortages, demography, destruction of property, and many other socioeconomic factors. To apprehend the rainfall evolution, this paper presents a statistical model which involves the lagged oceanic SOI. The objective is to characterize the seasonal dependence structure and to use it for forecasting rainfalls. The study confirms that the structure between Rainfall and SOI is strong on the central part while rainfall extreme events might be related to other atmospheric and climatologic indices. Thus, the study highlights the need of using a multivariate copula approach, given that rainfall are correlated with several atmospheric, geographic and environmental conditions.

Appendix

Given two random variables X and Y with continuous marginals F and G, Sklar's theorem (Sklar [22]) states that the joint distribution function $H(.,.)$ of (X, Y) can be written in terms of a unique function $C(F(.), G(.))$:

$$H(x, y) = C(F(x), G(y)) = C(u, v) \tag{1}$$

The function $C(.,.)$ is known as the copula function of (X, Y). It describes how H is coupled with the marginal functions F and G. The books by Nelsen [23] and Joe [24] are very good references on copula theory. For reader convenience, we present below some copula families we use in this work (Table 2).

Elliptical Copulas: They are very appropriate to capture symmetric dependence. Gaussian and t-copulas are famous examples of elliptical copulas. The main advantage of a t-copula is that it can capture tail dependence.

Archimedean Copulas: They are adequate in the case of tails dependence. The Clayton, Frank and Gumbel copulas are Archimedean. The Frank copula is

Table 2. Some examples of copulas

Copulas	θ	$C(u,v)$
Gumbel	$[1,\infty)$	$\exp[-((-\log(u))^\theta + (-\log(v))^\theta)^{\frac{1}{\theta}}]$
Clayton	$\theta \in [-1,\infty)\backslash\{0\}$	$\max([u^{-\theta} + v^{-\theta} - 1]^{-\frac{1}{\theta}}, 0)$
Frank	$\theta \in \mathbb{R}\backslash\{0\}$	$-\frac{1}{\theta}\ln(1 + \frac{(\exp(-\theta u)-1)(\exp(-\theta v)-1)}{\exp(-\theta)-1})$
Gaussian	$[-1,1]$	$\Phi_\Sigma(\Phi^{-1}(u), \Phi^{-1}(v))$
Student	$[-1,1]$	$t_{\Sigma,\nu}(t_\nu^{-1}(u), t_\nu^{-1}(v))$

symmetric while Clayton and Gumbel are asymmetric. As the Clayton copula exhibits greater dependence in the lower tail, the Gumbel copula captures upper tail dependence.

Once the copula $C(.,.)$ modeling the dependence between variables X and Y is identified, forecasting becomes a straightforward operation. For instance, using the conditional copula C_X, we can forecast Y given X. The conditional copula C_X is defined as:

$$P[Y \leq y|X = x] = P[V \leq v|U = u] = \frac{\partial C(u,v)}{\partial u} := C_X(F(x), G(Y)), \quad (2)$$

where $V = G(Y)$, $v = G(y)$, $U = F(X)$, and $u = F(x)$.

For $\tau \in [0,1]$, the τ-th conditional quantile function given $X = x$, denoted by $Q_X(\tau|x)$, is defined by:

$$Q_X(\tau|x) = G^{-1}(C_X^{-1}(F(x), \tau)), \quad (3)$$

where φ^{-1} denotes the inverse function of a function φ.

In this paper, we set τ to 0.5 which corresponds to the case of the median quantile Copula.

References

1. American Meteorological Society (AMS), Boston, Massachusetts, United States. http://www.ametsoc.org
2. Chawla, L., Osuri, K.K., Mujumdar, P.P., Niyogi, D.: Assessment of the Weather Research and Forecasting (WRF) model for simulation of extreme rainfall events in the upper Ganga Basin. Hydrol. Earth Syst. Sci. **22**(2), 1095 (2018)
3. Pham, M.T., Vernieuwe, H., De Baets, B., Verhoest, N.E.: A coupled stochastic rainfall-evapotranspiration model for hydrological impact analysis. Hydrol. Earth Syst. Sci. **22**(2), 1263 (2018)
4. Jha, S.L., Shrestha, D.L., Stadnyk, T.A., Coulibaly, P.: Evaluation of ensemble precipitation forecasts generated through post-processing in a Canadian catchment. Hydrol. Earth Syst. Sci. **22**(3), 1957 (2018)
5. Zhang, S., Zhao, L., Delgado-Tellez, R., Bao, H.: A physics-based probabilistic forecasting model for rainfall-induced shallow landslides at regional scale. Nat. Hazards Earth Syst. Sci. **18**(3), 969 (2018)

6. Kim, K., Lee, S., Jin, Y.: Forecasting quarterly inflow to reservoirs combining a copula-based bayesian network method with drought forecasting. Water **10**(2), 233 (2018)
7. Sene, K., Tych, W., Beven, K.: Exploratory studies into seasonal flow forecasting potential for large lakes. Hydrol. Earth Syst. Sci. **22**(1), 127 (2018)
8. Chiew, F.H., Piechota, T.C., Dracup, J.A., McMahon, T.A.: El Nino/Southern Oscillation and Australian rainfall, stream flow and drought: links and potential for forecasting. J. Hydrol. **204**(1–4), 138–149 (1998)
9. McBride, J.L., Nicholls, N.: Seasonal relationships between Australian rainfall and the Southern Oscillation. Mon. Weather Rev. **111**(10), 1998–2004 (1983)
10. Stone, R., Hammer, G., Nicholls, N.: Frost in northeast Australia: trends and influences of phases of the Southern Oscillation. J. Clim. **9**(8), 1896–1909 (1996)
11. Kirono, D.G., Chiew, F.H., Kent, D.M.: Identification of best predictors for forecasting seasonal rainfall and runoff in Australia. Hydrol. Process. **24**(10), 1237–1247 (2010)
12. Hyden, L., Sekoli, T.: Possibilities to forecast early summer rainfall in the Lesotho Lowlands from the El-Nino/Southern Oscillation. WATERSA-PRETORIA **26**(1), 83–90 (2000)
13. Hadiani, M., Asl, S.J., Banafsheh, M.R., Dinpajouh, Y., Yasari, E.: Investigation the Southern Oscillation index effect on dry/wet periods in north of Iran. Int. J. Agric. CropSci **4**, 1291–1299 (2012)
14. Chifurira, R., Chikobvu, D.: A probit regression model approach for predicting drought probabilities in Zimbabwe using the Southern Oscillation Index. Mediterr. J. Soc. Sci. **5**(20), 656 (2014)
15. Rodriguez, D., de Voil, P., Hudson, D., Brown, J.N., Hayman, P., Marrou, H., Meinke, H.: Predicting optimum crop designs using crop models and seasonal climate forecasts. Sci. Rep. **8**(1), 2231 (2018)
16. Zahmatkesh, Z., Goharian, E.: Comparing machine learning and decision making approaches to forecast long lead monthly rainfall: the city of vancouver, Canada. Hydrology **5**(1), 10 (2018)
17. Agilan, V., Umamahesh, N.V.: El Niño Southern Oscillation cycle indicator for modeling extreme rainfall intensity over India. EcologicalIndicators **84**, 450–458 (2018)
18. Deheuvels, P.: La fonction de dépendance empirique et ses propriétés. Un test non paramétrique d'indépendence. Acad. Roy. Belg. Bull. Cl. Sci. **65**(6), 274–292 (1979). 5e serie
19. Roncalli, T., Durrleman, A., Nikeghbali, A.: Which copula is the right one. Groupe de Recherche Opérationnelle, Credit Lyonnais, Paris (2000)
20. Embrechts, P., Lindskog, F., McNeil, A.: Modelling dependence with copulas. Rapport technique, Département de mathématiques, Institut Fédéralde Technologie de Zurich, Zurich (2001)
21. Kolev, N., Paiva, D.: Copula-based regression models. Department of Statistics, University of São Paulo (2007)
22. Sklar, A.: Fonctions de répartition à n dimension et leurs marges. Publications de l'Institut Statistique de l'université de Paris **8**, 229–231 (1959)
23. Nelsen, R.B.: An Introduction to Copulas, 2nd edn. Springer, New York (2006)
24. Joe, H.: Multivariate Models and Dependence Concepts. Monographs on Statistics and Applied Probability, vol. 73. Chapman Hall, London (1997)

Toward a Study of Environmental Impact of Shipping Industry and Proposal of Alternative Solutions

Abdelmoula Ait Allal[✉], Khalifa Mansouri, Mohamed Youssfi,
and Mohammed Qbadou

Laboratory: Signals, Distributed Systems and Artificial Intelligence (SSDIA),
ENSET Mohammedia, University Hassan II, Casablanca, Morocco
aitallal.abdelmoula67@gmail.com,
khmansouri@hotmail.com, med@youssfi.net,
qbmeden7@gmail.com

Abstract. The maritime industry is considered as the main pillar for the worldwide economy trade. It is the propulsion engine for the economy development and prosperity of many countries. This shipping industry encounter many challenges to ensure its competitiveness and its environmental sustainability. However, the ship operation generates many environmental pollutants that harm the air and oceans. It daily produce sludge, bilge water, garbage, sewage and harmful the ecosystem by the ballast water exchange operation. In this paper, we aim at identifying and quantifying these products and to assess their impact on the marine environment. Alternative solutions are proposed to reduce the shipping environmental impact. The concept of autonomous ship is also proposed as an alternative for an environmentally sustainable maritime industry.

Keywords: Shipping industry · Ocean pollution · Environment · Conventional ship · GHG emission

1 Introduction

The maritime industry participates with 12% in the total sea pollution [1]. The way the maritime industry responds to the environmental problems may in fact be a leading indicator of its overall competitiveness. Most the trading ships use Heavy fuel oil as propulsion energy. The use of this type of fuel oil leads to the production of sludge, greenhouse gases (GHG) emission, and in case of oil spill may cause catastrophic harmful to the sea. The presence on board of the crew and daily maintenance on board produce garbage, sewage and increase the energy consumption. In addition, for its safe operation, the ship needs water ballast exchange operation to control its list, trim, stability, and structures stress. the water ballast exchange may cause immigration of unwanted species and pathogens from one area to another leading to an unbalanced ecosystem. To reduce the environmental impact of the conventional ships, the international maritime organization has established several conventions, i.e. Marpol annex I, which deal with oil pollution prevention [2], Marpol annex IV, which deal with

© Springer Nature Switzerland AG 2019
M. Ezziyani (Ed.): AI2SD 2018, AISC 913, pp. 245–256, 2019.
https://doi.org/10.1007/978-3-030-11881-5_21

sewage pollution prevention [3], Marpol annex V, which deals with garbage pollution prevention [4], Marpol annex VI, which deal with air pollution prevention [5]. and the international water ballast management system for the management and records of the water ballast movements [6]. Our approach to this study consists of the visit of several ships of type, general cargo ship, bulk carrier ship and container ships. The analysis on board of the relevant documents to identify and quantify the produced environmental pollutant products and to assess the implemented measures to reduce their harmful character. In order to enhance the ocean pollution prevention, we propose alternative environment friendly solutions, i.e. use of clean energy, and implementation of autonomous ship (AS) in the maritime industry. Nowadays, many autonomous ship concepts have been proposed, i.e. Rolls-Royce with its partners developed a vision of remote controlled ship operation and the implementation of AS in the maritime industry [7]. In the maritime unmanned navigation through intelligence in network (MUNIN) project, the feasibility and conception of AS have been studied [8–10]. The classification society DNVGL proposed a concept called "REVOLT" using electrical propulsion [11].

The ocean pollution by shipping has been a subject of many studies and literatures. In [12], the authors aimed at quantifying the impact of the ports implemented effort regarding the reduction of GHG emission. The different measure to reduce the emission were investigated. Three scenarios were proposed for the 2030 GHG emission reduction in the port of Gothenburg, i.e. alternative fuel, ship design, and operation. The scenario analysis identifies difficulties to reach significant reductions of GHG emissions in the studied port by 2030. The biggest challenge is the reduction of ships GHG emission when they are in the port. In [13], the authors suggest associating damages and policies with ports, emission control areas (ECA), and open seas. In their study, they argued to burn dirty fuel oil at high seas and quantified the benefits and cost saving from reforming current IMO and other approaches towards environmental management with three-layer approach. The study resulted in that the international maritime organization (IMO) and other authorities should reconsider their decision of allowable Sulphur content in fuel oil from 3.5% to 0.5% by 2020. With the aim at reducing the illegal discharge of ships waste at sea, the European Community has obliged the vessels calling the European port to pay a single tariff whether they discharge waste or not. In majority of European ports there is a single tariff, based on vessel size, includes the discharge oil waste and garbage. In [14], the authors aimed at determining if other factors also affect waste generation. The results shew that the main driver in the generation of oily waste is ship size while the main factor in the generation of garbage is people on board of ships. These results point to the necessity to adopt different tariff for each type of waste. In [15], the authors have studied the effects of the fraction that ends up in the water column and to which aquatic and sediment-dwelling organisms are exposed and carried out an integrated environmental risk assessment of the effects of emissions from oceangoing ships including the aquatic. Their Research focused on the quantitative and qualitative determination of pollutant emissions from ships and their distribution and fate, including the in-situ measurement of emissions in ships in order to derive realistic emission factors, and the application of atmospheric and oceanographic transportation and chemistry models.

The present paper is organized as follow: in Sect. 1, we introduce the importance of the studied thematic, presenting the maritime challenges to ensure its sustainability. The related literatures are cited. In Sect. 2, the used materials and methods to collect data are presented. In Sect. 3, the use of fuel oil (FO) for the ship propulsion is assessed in term of environmental impact on the oceans. In Sect. 4, the environmental impact of human presence onboard of conventional ship (CS) is studied. In Sect. 5, the water ballast exchange impact on ecosystem is studied and alternative innovative ship designs are presented to reduce the immigration of aquatic species. In Sect. 6, we present the implementation of autonomous ship (AS) as an alternative for the ocean's pollution prevention. In Sect. 7, we conclude our paper.

2 Data Collection and Approach Methodology

For an accurate identification and quantification of the ocean shipping pollutants products, several ships have been attended for an in-situ study. The relevant documents are deeply studied to extract the pertinent information. The following documents are scrutinized for a deep assessment,

- Oil records book: where the chief Engineer records the onboard retained oil waste and bilge water, and where he records also, the bunkered fuel oil (FO), marine Gaz oil (GO) and lubrication oil (LO) quantities. The chief engineer are requested to report in the same book the discharged quantity of oil sludge and bilge water to the ports facilities. When the oily water separator is used to separate the oil from bilge water and to discharge the separated water with less than 15 ppm overboard. The treatment date, start, stop geographic position and quantity, must be reported.
- FO, GO, and LO bunkering delivery notes folder: where we can find the FO, and GO, received quantities and its Sulphur content.
- FO laboratory analysis reports: where we can find the FO analysis data and the recommended heating temperature for its burning.
- Engine log book: Where daily engine room parameters, maintenance operation, voyage data are recorded.
- Deck log book: where navigation data, voyage data are recorded.
- Garbage management plan, garbage record book and garbage delivery notes: where the chief officer report the retained quantity of different types of garbage, including plastic, food waste, domestic wastes (Paper products, rags, glass, metal, bottles, crockery, etc.), cooking oil, incinerator ashes, operational wastes, cargo residues, animal carcasses, fishing gear. The chief officer is requested also to report the quantity delivered to the port facility and to keep the delivery notes as evidence.
- Water ballast management plan and water ballast movement log book: Where the chief officer records the water ballast movement including date, identification of the ballast tank, start, stop geographic position, quantity of water transferred.

- Ozone depleting substances record book: where the chief engineer records the ozone depleting substances. These substances contain usually chlorine or bromine. They usually found as cooling agents or agents in fire-fighting equipment in older ships. The chief engineer shall made entries in the record book for any recharge, repair maintenance of equipment containing ozone depleting substances, discharge to atmosphere (deliberate or non-deliberate), discharge to land-based reception facilities, and supply of ozone depleting substances to the ship.
- addition other documents such as, noon reports, sea passage reports, port state control (PSC) report, environment incident reports, environment nonconformities reports, are studied. The ship's crew members are interviewed for their experienced cases and their feedback regarding the environmental issues and the way they manage the garbage, sewage, water ballast movement onboard.

The collected data show the considerable environmental impact of the shipping industry on the ocean and urge the shipping players to think differently at the ship design phase and at the operation phase. The ship-owners must implement innovative technologies and efficient management systems.

3   Environmental Impact of the Use of FO for the Ship Propulsion

Nowadays the most of the ships use FO and DO for their propulsion. This type of propulsion energy generates a considerable amount of oil sludge, bilge water, and greenhouse gases (GHG) emission, and in case of oil spill may cause catastrophic harmful to the environment.

3.1   Shipping Oil Sludge and Bilge Water Production

The use of FO onboard of conventional ships lead to the production of sludge and bilge water. The FO is used for the running of the main engine, auxiliary engines and the boiler. Before to be used, the FO oil must be purified through a purification process and heated before its entry in the engine injection system. This process consumes electrical energy and generates great amount of sludge. This sludge must be retained onboard and discharged to the land-based facilities. Unfortunately, it is not all the time the case. According to our interview, an uncontrollable amount of oil residues is still be discharged overboard by lack of awareness and to minimize the ship operation cost. These discharged residues harmful the ocean and cause its pollution and destruction of life at sea. As case study, we take a ship of type container ship,

Ship type: Container ship
Length overall: 187 m
Propulsion power: 16660 KW (1 main engine)
Auxiliary engine power: 1400 KW X3 (3 auxiliary engines)
Gross tonnage: 18327 Ton
Crew number: 20

In this case study, the FO used is of type RMK 500, with an average consumption of 69.1 MT per day and about 24185 Ton per year with a Sulphur content of 2.4% (Table 1). The use of this type of FO generates 0.15 MT per day, 52.5 per year of sludge and 1.2 MT per day and 420 MT per year of bilge water (Table 2). In compliance with Marpol Annex I and based on the delivery notes, the recorded amount of sludge and bilge water in the oily record book is discharged to land-based facilities. To avoid penalties and detention of the ship at port, the chief engineer makes entries in such a way that the produced residues amount and the delivered amount are compatible. But for sure is not reflecting the reality because during my survey the purification process is not good maintained and lot of FO leakages are noted, which demonstrate that the real generated amount is much bigger than what is reported. This uncontrolled discharge of residues, destroy the life at sea.

Table 1. FO consumption and Sulphur content.

Equipment	Fuel type	Sulphur content	Daily consumption (MT)	Yearly consumption (MT)
Main engine	FO	2.4	59.00	20300
Auxiliary engines	FO	2.4	9.50	3325
Boiler	FO	2.4	1.60	560

Table 2. Oil sludge and bilge water production on board of conventional ship.

Liquid type	Monthly production (MT)	Yearly production (MT)
Sludge	0.15	52.5
Bilge water	1.2	420

3.2 Environmental Impact of GHG Emission on Oceans

The concern and interest of the maritime players about atmospheric inputs to the ocean has grown significantly, in particular, in the major shipping routes where the impacts of GHG emission is expected to be greater. The shipping industry participate with 3% in the global GHG emission [16]. The emission of various pollutants which are derived from ship exhaust, i.e. SO_2, NO_x, and particulates. The input from the atmosphere of these shipping pollutants increase the acidification of the ocean due to rising level of CO_2. Greater the sea water is acidic less its natural CO2 sink that it provides is maintained. The absorption of gases such as Sulphur and nitric acids lead to the decrease of sea water pH, which in turn cause the release of CO2 from the ocean to the atmosphere [17] (Fig. 1). The GHG emission depends on the vessel commercial trips frequency, the type of FO used and the running condition of the machinery. In order to reduce these GHG, and to comply with Marpol annex VI, the ship shall use FO with less than 3.5% Sulphur content, and while she is operating within an emission control area (ECA), the Sulphur content of the used FO shall not exceed the limit of 0.5%. The change over from high Sulphur FO to low Sulphur FO in these ECA's has caused several main engine stoppages, leading to many catastrophic maritime accidents. To reduce the FO negative environmental impact, an alternative clean energy should be used, i.e. DO, distillated FO, liquefied natural gas (LNG), and electrical propulsion.

Fig. 1. Acidification of oceans by GHG emission.

4   Environmental Impact of the Human Presence on Board

The presence of human onboard generates various ocean pollutants, i.e. garbage, sewage, and increase of the GHG emission.

4.1   Garbage Production on Board of Conventional Ship

The conventional ship (CS) produces various types of garbage, i.e. plastic, food waste, domestic wastes (Paper products, rags, glass, metal, bottles, crockery, etc.), cooking oil, incinerator ashes, operational wastes, cargo residues, electrical waste, animal carcasses.

The produced and discharged amount is recorded accordingly in the garbage record book. This production of garbage is due to the human presence on board and to the machinery routine maintenance. In our case study ship, the crew produce about 7 M^3 per day, 85 M^3 per year and the routine maintenance on deck and in the engine room produces about 1.8 M^3 per day, 19.5 M^3 per year (Table 3). The produced amount of garbage has been found recorded in the garbage record book. Unfortunately, the recorded amount does not reflect the reality. Because the number of crew on board and the excessive maintenance onboard surely result in the production of an amount of garbage bigger that the one recorded. Thus, a part of the garbage is thrown overboard to sea during the sea passage to reduce the ship operation cost. This garbage amount might be reduced by clean propulsion energy and less crew number. The only way to motivate the crew to keep the garbage on board and to deliver it to the land-based facilities is to be free of charge and that governments take in charge the treatment cost.

Table 3. Garbage production on board of conventional ship.

Garbage type	Monthly production (M^3)	Yearly production (M^3)
Produced by crew	7.00	85.00
Produced by engine room	1.80	19.500

4.2   Sewage Production on Board of Conventional Ship

The presence of human on board of CS leads to the production of sewage i.e. black water and grey water which are treated in sewage plant or collected in a holding tank. The black water is the sewage that is produced by using toilets, whereas, the grey water is the sewage, that is generated from the galley, laundry, showers and cleaning. The sewage which is generated from the vessel contains higher concentration of Bio-chemical Oxygen Demand (BOD) and Suspended Solid (SS) than sewage from the land. In compliance with Marpol 73/78, Annex IV, this sewage must be treated before be discharged overboard to the sea by the sewage treatment plant to make it less environmentally harmful. The IMO recommends that the treated sewage should be with a BOD, SS less than 50 mg/l and a coliform count less than 250 Fecal per 100 ml. In our case study, the vessel generates about 5 MT per day, 1750 MT/year of grey water and 0.06 MT per day, 21 MT per year (Table 4). The low production of the black water is due to that the toilets are equipped with low water consumer flush system. The treatment system must operate continuously without stoppage. Unfortunately, is not all the time the case. Based on the interview of the crew, the sewage treatment plant most of the time was kept stopped, and the sewage was discharged overboard without treatment. With less crew members and implementation of autonomous ship (AS), the production of sewage might be reduced and thereby the ocean pollution can be prevented.

Table 4. Sewage production on board of conventional ship.

Sewage type	Monthly production (MT)	Yearly production (MT)
Grey water	5.00	1750.00
Black water	0.06	21.00

4.3   GHG Emission

The presence of human onboard necessitates various facilities for his human life being and comfort. These facilities include, lighting, sewage treatment plant, air conditioning, galley, laundry and provision reefer plant. On board the crew consume about 10807.2 KWH per day (Table 6). This energy is produced by the auxiliary engine, that uses FO for its running. This contributes considerably in the GHG emission and thereby in the ocean pollution. This emission can be reduced by less crew members and implementation of autonomous ship (AS).

5   Ballast Water Exchange Environmental Impact

To control the ship trim, list, draught, stability, and structure stress under all loading conditions, the crew carry water ballast transfer. The Table 5 presents the amount of ballast water exchanged on board of our case study. The ballast pumping system consumes a considerable energy for the running of the associated ballast pumps. The transfer of ballast water from one area to another presents a potential risk of immigration of aquatic species, leading to an unbalanced ecosystem. These can be avoided by adoption of new concepts of hull design. This concept consists of non-ballast water ship (NOBS) concept, which is based on V-shaped hull design. The Det Norse Veritas (DNV) and Delft University of Technology (DUT) hull design for zero ballast is a new thinking about optimal hull shape and buoyancy distribution, and it represents another major design change away from the flat-bottom hull of the conventional trading ships (Fig. 2).

Ballast free ship using continuous flow method   Zero-ballast concept using V-shaped hull

Fig. 2. Non-ballast ship design concept.

These "alternative" methods fail in two categories, "non-ballast/zero discharge" methods, and "continuous flow" methods. The category of "non-ballast" includes the ships with novel hull design [18] and others ranging from those that carry ballast temporarily (e.g. during storms), to those carrying freshwater that is shifted from tank to tank and seldom discharged, to those that carry drinking water that can be discharged in port if necessary. The "continuous flow" methods embrace various engineering concepts including the replacement of ballast tanks by longitudinal trunks, modified free flow ballast tanks and tanks that allow for enhanced exchange by pumping in the open ocean. With the non-ballast water ship concepts, there is no longer needs in the installation of the ballast pumps and ballasting system piping and valves, which result in the decrease of the light ship weight, reduction of energy consumption, and GHG's emission, and mitigate the immigration of unwanted species and pathogens from one area to another.

Table 5. Sewage production on board of conventional ship.

Movements	Monthly exchanged amount (M^3)	Yearly exchanged amount (M^3)
Exchanged amount of ballast water	5.00	1750.00

6 Oceans Pollution Prevention by Implementation of Autonomous Ship

The concept of autonomous ship consists of the elimination of crew onboard and to be remotely controlled and monitored from the shore control center based on shore. It is assumed that the AS will use clean energy for its propulsion such as liquefied natural gas (LNG), distilled FO, GO, and electrical propulsion. In addition, the AS design will be with non-ballast design. The implementation of autonomous ship will have a great impact on energy saving and reduction GHG's emission. As the autonomous ship, will operate without crew onboard, the accommodation and crew life being facilities are no longer needed. The elimination of crew onboard will reduce considerably the energy consumption and production of garbage and sewage. The non-ballast design will eliminate the ballasting system related consumers (pumps, control systems). The Table 6 shows the impact of CS and AS in term of the production of ocean pollutants. The sign (++) means greater impact and the sign (–) means less impact. The Table 7 depicts the equipment to be eliminated and its resulting in energy-saving.

Table 6. Autonomous ship implementation impact on oceans pollution.

Pollutant type	Conventional ship	Autonomous ship
GHG emission	++	–
Garbage	++	–
Sewage	++	–
Water ballast exchange	++	–

Table 7. Impact of human elimination on energy consumption on board of autonomous ship.

Consumers	Nominal power (KW)	Operation at sea nominal power			
		Nominal power (KW)	CS (KW)	AS (KW)	Expected reduction (%)
Inner lighting	32.70	26.20	26.20	13.10	50
Outdoor lighting	18.80	15.00	15.00	7.50	50
Sewage treatment plant	8.30	7.60	7.60	0	100
Shower water heating	32.60	32.30	32.30	0	100
Ballast pump	250.00	239.40	239.40	0	100
Ballast water	109.00	92.70	92.70	0	100
Air conditioning compressor	127.00	63.50	63.50	0	100
Air conditioning supply fan	35.00	30.80	30.80	0	100
Air conditioning heating	212.30	210.20	210.20	0	100
Air condition. for air bridge	5.60	5.10	5.10	5.10	0
Air conditioning for galley	4.40	4.30	4.30	0	100
Provision reefer unit	16.90	14.90	14.90	0	100
Provision reefer store fan	0.20	0.20	0.20	0	100
Laundry/drying room fan	0.10	0.10	0.10	0	100
Sanitary space exhaust fan	0.10	0.10	0.10	0.10	0
Galley exhaust fan	1.30	1.30	1.30	0	100
Galley supply fan	0.70	0.70	0.70	0	100
Hospital exhaust fan	0.20	0.20	0.20	0	100
Cooking range	15.00	9.00	9.00	0	100
Baking hoven	9.00	3.60	3.60	0	100
Deep fryer	6.00	4.80	4.80	0	100
Refrigerators	12.00	5.00	5.00	0	100
Dish washing machine	10.60	8.00	8.00	0	100
Washing machine	32.4	19.4	19.4	0	100
Drying machine	9.20	3.70	3.70	0	100
Electric iron	8.40	5.00	5.00	0	100
Total	957.8	803	803	25.8	96.84

7 Conclusion

The ocean pollution results in threats to marine ecosystems, human health and diminishing of food sources. Our study shows that the conventional ships have various harmful impacts on oceans, i.e. pollution of the sea by oil spill, and uncontrolled discharge of sewage and garbage, risk of unbalanced marine ecosystem by water ballast movements from one area to another, and pollution of air by greenhouse gases emission. These ocean pollutants may be reduced by using clean energy for the propulsion of the conventional ships and to implement innovative technology, i.e. enhanced automation and internet of thing (IoT) technologies. One of the IoT technology is the future implementation of autonomous ship concept. The autonomous ship concept still in its infancy phase but might be considered as an alternative for the implementation of an environmental friendly maritime industry.

References

1. Group of Experts on the Scientific Aspects of Marine Pollution (GESAMP), Pollution in the open oceans (2015). http://www.gesamp.org
2. International Maritime Organization (IMO), Marpol 73/78, Annex I, Regulations for the prevention of pollution by oil (2016)
3. International Maritime Organization (IMO), Marpol 73/78, Annex IV, Regulations for the prevention of pollution by sewage from ships (2016)
4. International Maritime Organization (IMO), Marpol 73/78, Annex V, Regulations for the prevention of pollution by garbage from ships (2016)
5. International Maritime Organization (IMO), Marpol 73/78, Annex VI, Regulations for the prevention of air pollution from ships (2016)
6. International Maritime Organization (IMO). Ballast Water Management (2004)
7. Rolls-Royce Plc. Remote and autonomous ships. Advanced Autonomous Waterborne Applications (AAWA) partners (2016). www.rools-royce.com/marine
8. Kretschmann, L., Rodseth, O., Tjora, A.: Report D9.2: Qualitative assessment, Maritime Unmanned Navigation through Intelligence in Networks (MUNIN) (2015)
9. Kretschmann, L., Rodseth, O., Fuller, B.S.: Report D9.3: Quantitative assessment (MUNIN) (2015)
10. Rodseth, O., Burmeister, H.C.: Report D10.2: New ship designs for autonomous ship (MUNIN) (2015)
11. DNVGL, ReVolt-next generation short sea shipping (2017). https://www.dnvgl.com/news/revolt-next-generation-short-sea-shipping-7279
12. Winnes, H., Styhre, L., Fridell, E.: Reducing GHG emission from ships in port area. J. Res. Transp. Bus. Manage. **17**, 73–82 (2015)
13. Lindstad, H., Eskeland, G.S., Psaraftis, H., Sandaas, I., Strømman, A.H.: Maritime shipping and emissions: a three-layered, damage-based approach. J. Ocean Eng. **110**, 94–101 (2015)
14. Pérez, I., González, M.M., Jiménez, J.L.: Size matters? Evaluating the drivers of waste from ships at ports in Europe. J. Transp. Res. Part D **57**, 403–412 (2017)
15. Blasco, J., Durán-Grados, V., Hampel, M., Moreno-Gutiérrez, J.: Towards an integrated environment risk assessment of emission from ship's propulsion systems. J. Environ. Int. **66**, 44–47 (2014)

16. International maritime organization (IMO). Second IMO GHG study 2009, London (2009). http://www.imo.org/
17. Group of Experts on the Scientific Aspects of Marine Polution (GESAMP). The atmospheric input of chemicals to the ocean, GSAMP reports and studies, no. 84 (2012)
18. GEF, UNDP, IMO, GloBallast Partnerships, GESAMP. Establishing Equivalency in the Performance Testing and Compliance Monitoring of Emerging Alternative Ballast Water Management Systems (EABWMS). A Technical Review (2011)

Optimization of Coagulation Flocculation Process for the Removal of Heavy Metals from Real Textile Wastewater

Dalila Sakhi[1(✉)], Younes Rakhila[1], Abedellah Elmchaouri[1],
Meriem Abouri[2], Salah Souabi[2], and Amane Jada[3]

[1] Laboratory of Physical and Bioorganic Chemistry,
Faculty of Sciences and Technics of Mohammedia,
Hassan II University, Mohammedia, Morocco
sakhi.dalila@gmail.com
[2] Laboratory of Process Engineering and Environment,
Faculty of Sciences and Technics of Mohammedia,
Hassan II University, Mohammedia, Morocco
[3] Institute of Materials Sciences of Mulhouse (IS2M-UMR 7361 CNRS - UHA),
Mulhouse, France

Abstract. The coagulation flocculation process with ferric chloride as coagulant and polymer as a flocculant was optimized for the elimination of heavy metals from a real textile wastewater using 4^2 composite central and surface response method. The effect of the three factors (pH, dose of coagulant and volume of flocculant) on the elimination of heavy metals was investigated and found to be positive. The optimal conditions obtained from the compromise of the desirable responses such as Cd removal, Pb removal, As removal, Ni removal and Se removal were 0.64 g/L of coagulant dosage, 2.6 g/L of flocculant dosage at pH 8.1. The maximum removal of Cd, Pb, Ni, As and Se in this study achieve respectively 38.49%, 78, 88% and 61.88% 60.63%, 81.76%, 47.01% in optimal conditions.

Keywords: Real textile wastewater · Coagulation flocculation · Heavy metals · Optimization · Experimental design · Surface response methodology

1 Introduction

Nowadays, wastewater is one of the most critical environmental issues faced by countries around the world. One of the most hazardous effluents is the textile wastewater because they are highly charged with organic and mineral matter. Besides colour visibility which brings displeasing aesthetics, heavy metal constituents in the effluent also resulted in negative ecological impacts to the water-body, environment as well as deterioration of human health [1]. Which requires treatment before discharge to the receiving environment.

The methods used for contaminant removals from dye wastewater can be divided into three main categories; physical, chemical and biological. Physical treatments such as precipitation, ion exchange, membrane filtration, irradiation, ozonation and

© Springer Nature Switzerland AG 2019
M. Ezziyyani (Ed.): AI2SD 2018, AISC 913, pp. 257–266, 2019.
https://doi.org/10.1007/978-3-030-11881-5_22

adsorption are widely used technics [2]. Physico-chemical treatment methods are coagulation flocculation, precipitation, photo-catalysis, oxidation and chemical sludge oxidation lastly, biological treatment technics used are aerobic degradation, anaerobic degradation [3].

Coagulation flocculation is one of the most encouraging processes that can be used as a physico-chemical method in the treatment of metal bearing industrial wastewater because it removes [4, 5] colloidal particles, some soluble compounds, and very fine solid suspensions initially present in the wastewater by destabilization and formation of flocs. Both aluminum and ferric salts, either in monomer or polymeric forms, have been reported as effective coagulants in treating heavy metals from wastewater [6].

Coagulation–flocculation is commonly used for textile wastewater, because of its efficiency and simplicity [7, 8]. This efficiency depend on many factors such as the type and dosage of coagulant/flocculant and pH [9] and can be increased by optimizing these factors. In conventional multifactor experiments, optimization is usually carried out by varying a single factor while keeping all other factors fixed at a specific set of conditions. It is not only time-consuming, but also usually incapable of reaching the true optimum due to ignoring the interactions among variables. Response surface methodology (RSM) has been proposed to determine the influences of individual factors and their interactions [10].

This study is dedicated to the optimization of the operating conditions such as pH, coagulant dose and flocculant volume for the treatment of real wastewater coming from a textile industry located in Casablanca, Morocco; by coagulation flocculation processes with ferric chloride (40%) as coagulant and polymer (polyacrylamide Himoloc DR3000) as a flocculant using the methodology of the experimental designs based on the response surface methodology (RSM) to optimize and improve the removal efficiency of Cd, Pb, Cr, Ni, As and Se.

2   Materials and Methods

2.1   A Sample Collection

The sample was taking from a textile industry located in Ain Sbaa, Casablanca, Morocco. The wastewater textile was collected in 50 L plastic bottles, then transported to the laboratory and finally stored at 4 °C. These samples were then removed from the refrigerator and left at ambient temperature for about 2 h before their characterization and their treatment.

2.2   Chemicals and Materials

The coagulant used is ferric chloride 40%. The flocculant used is an anionic polymer 35%; its trade name is Himoloc DR3000.

The experimental set-up used for the coagulation–flocculation experiments at laboratory scale consisted of a Jar-test device (Jar Test Flocculator FC-6S Velp Scientific) in which six stirring blades were connected to a motor that operated under adjustable conditions. The system permitted the experiments to be performed with ease and the different variables affecting the removal of suspended fat and organic matter to be interpreted such as pH, stirring time and speed, retention time or reactant concentrations.

2.3   Analytical Procedure

The pH of textile wastewater was adjusted by addition of NaOH (40%) to a desired value in the range of 6.42–9.78. Coagulant dosages (FeCl3) varied in the range of 0.37–0.9 mg/L, while flocculant dosages (Himoloc DR3000 0.3%) ranged from 0.9 to 4.3 mL/L. Sixteen experiments were carried out. After the addition of coagulant, the textile wastewater was stirred at 300 rpm for 10 min. Then the flocculant was added and stirred at 30 rpm for 20 min. Samples were taken from the supernatant and analysed after leaving the medium to stand for two hours.

The samples of textile wastewater were measured using a turbidimeter (Model HACH 2100 N TURBIDIMETER). COD was determined by titrimetric method as described in standard methods [11] and BOD5 was determined by BOD meter. The heavy metal content was determined by the inductively coupled plasma technique (ICP). Removal efficiency of heavy metals were obtained according to the formula given below:

$$\text{Removal } (\%) = \left(1 - \frac{C}{C_0} \times 100\right) \tag{1}$$

Were C_0 and C are the initial and final concentration values.

2.4   Experimental Design

A central composite rotatable design for k independent variables was employed to design the experiments [12] which the variance of the predicted response, $\hat{Y}$, at some points of independent variables, X is only a function of the distance from the point to the design center. These designs consist of a 2 k factorial (coded to the usual ± 1 notation) augmented by 2*k axial points $(\pm\alpha, 0, 0)$, $(0, \pm\alpha, 0)$, $(0, 0, \pm\alpha)$, and 2 center points $(0, 0, 0)$ (13). The value of α for rotatability depends on the number of points in the factorial portion of the design, which is given in Eq. (2):

$$\alpha = (NF)1/4 \tag{2}$$

Where NF is the number of points in the cube portion of the design (NF = 2 k, k is the number of factors). Since there are three factors, the NF number is equal to 2^3 (=8) points, while α is equal to (8)1/4 (=1.682) according to Eq. (2). In this study, the responses were Pb removal (Y_{Pb}), Cr removal (Y_{Cr}), Ni removal (Y_{Ni}), As removal (Y_{As}), Se removal (Y_{Se}) and Cd removal (Y_{Cd}) of textile wastewater. Each response was used to develop an empirical model that correlated the response to the coagulation processes activated variables using a second- degree polynomial equation as given by Eq. (3):

$$\hat{Y} = \beta_0 + \beta_1 X_1 + \beta_2 X_2 + \beta_3 X_3 + \beta_{12} X_1 X_2 + \beta_{13} X_1 X_3 + \beta_{23} X_2 X_3 + \beta_{11} X_{12} + \beta_{22} X_{22} + \beta_{33} X_{32} \tag{3}$$

Where β_o the constant coefficient, β_i the linear coefficients, β_{ij} the cross-product coefficients and βii the quadratic coefficients. The software JMP PRO® 13 was used for the experimental design, data analysis, model building and graph plotting.

3   Results

3.1   Characterization of Textile Wastewater

The physicochemical characteristics of textile wastewater are shown in Table 1. It has a moderately basic pH, its turbidity reached 100 NTU, and represents a mineral pollution which is translated by hight value of Fe (μg/l) = 173 Cr (μg/l) = 9. Pb (μg/L) = 18. It is also loaded with organic matter, represented by a COD about 768 mg O_2/l these values exceed the Moroccan standards.

Table 1. Characterization of the textile effluent

Parameter	Value
pH	8.02
Turbidity (NTU)	100
BOD$_5$ (mgO$_2$/l)	24
Conductivity (ms/cm^{-3})	1.87
COD (mg O_2/l)	768
Cl (mg/l)	321
Na (mg/l)	178
Ca (mg/l)	284.62
NO$_3$ (mg/l)	51.25
NH$_3$ (mg/l)	3.01
NT (mg/L)	62.54
Mg (mg/l)	51
Fe (μg/l)	173
Pb (μg/L)	18
Ni (μg/l)	16
Cr (μg/l)	9

3.2   Development of the Regression Model Equation

Preliminary experiments were carried out to screen the appropriate parameters and to determine the experimental domain. From these experiments, the effects of pH of textile wastewater (X_1), coagulant dosage in mg/L (X_2) and flocculent dosage in mL/L (X_3) are investigated on 6 responses: Cd removal, Pb removal, Cr removal, Ni removal, As removal and Se removal. The parameter levels and coded values were given in Table 2.

Table 2. Study field and coded factors.

Natural variables (Xj)	Unit	Coded variables X1, X2, X3*				
		a	–	0	+	A
X1 = Initial pH	-	6.42	7.1	8.1	9.1	9.78
X2 = Coagulant dosage	mg/L	0.37	0.48	0.64	0.8	0.91
X3 = Floculant dosage	ml/L	0.92	1.6	2.6	3.6	4.28

* The coded values Xj = ±1 are obtained by equation: Xj = (xj – Xj)/Δ

The coefficient of determination R^2 in this study were relatively high, indicating a good agreement between the model predicted and the experimental values. Meanwhile, adjusted R^2 permitting for the degrees of freedom associated with the sums of the squares is also taken into account in the lack-of-fit test, which should be an approximate value of R^2. When R^2 and adjusted R^2 differ dramatically, there is a good chance that insignificant terms have been included in the model [14]. The experimental design matrix, the corresponding experimental parameters and response value were shown in Table 3.

Table 3. Experimental design and results for textile wastewater.

Tests	pH	FeCl$_3$ (mg/l)	Floc (ml/l)	Cd (%)	Pb (%)	Ni (%)	Cr (%)	As (%)	Se (%)
– – –	7.1	0.48	1.6	40	61,11	68,75	44,44	83,33	62,07
– – +	7.1	0.48	3.6	20	77,77	50	33,33	78,57	72,41
– + –	7.1	0.8	1.6	20	83,33	50	22,22	85,71	75,86
– + +	7.1	0.8	3.6	20	66,66	56,25	33,33	80,95	55,17
+ – –	9.1	0.48	1.6	10	72,22	50	33,33	76,19	41,38
+ – +	9.1	0.48	3.6	40	72,22	62,5	33,33	90,47	51,72
+ + –	9.1	0.8	1.6	30	77,77	62,5	44,44	85,71	75,86
+ + +	9.1	0.8	3.6	30	61,11	56,25	55,55	80,95	72,41
a00	6.4	0.64	2.6	30	77,77	56,25	33,33	80,95	58,62
A00	9.8	0.64	2.6	30	66,66	56,25	33,33	83,33	58,62
0a0	8.1	0.4	2.6	20	77,77	56,25	33,33	83,33	41,37
0A0	8.1	0.9	2.6	40	77,77	56,25	33,33	85,71	62,07
00a	8.1	0.64	0.9	30	77,77	56,25	55,55	85,71	55,17
00A	8.1	0.64	4.3	30	77,77	56,25	33,33	83,33	51,72
0	8.1	0.64	2.6	40	77,77	62,5	22,22	80,95	44,82
0	8.1	0.64	2.6	40	77,77	62,5	22,22	80,95	44,82

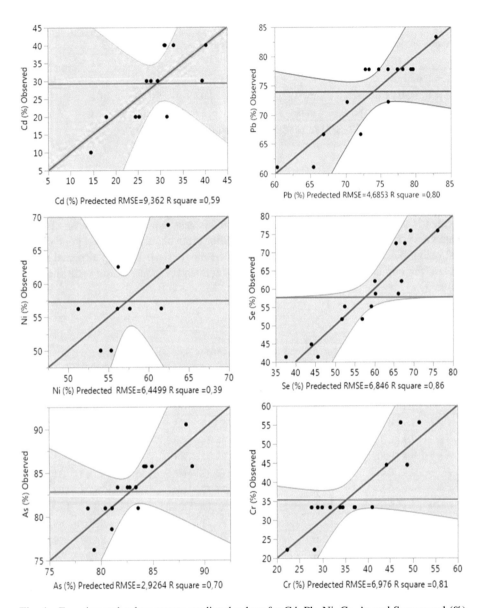

Fig. 1. Experimental values versus predicted values for Cd, Pb, Ni, Cr, As and Se removal (%)

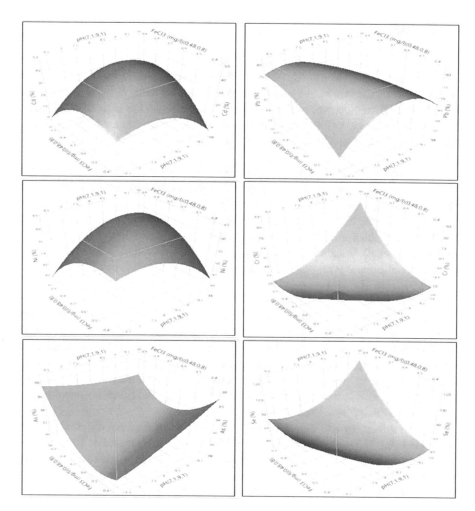

Fig. 2. Response surface plots for Cd, Pb, Ni, Cr, As and Se removal as a function of pH and coagulant dosage at flocculant dosage equal at 2

Table 4. Estimation of coefficients

Terme	Estimation	Standard error	t ratio	Prob. > |t|
Constant Cd	40.303301	6.600536	6.11	<0.0009*
Constant Pb	78.193791	3.303306	23.67	<0.0001*
Constant Ni	62.388037	4.547416	13.72	<0.0001*
Constant Cr	22.262086	4.918315	4.53	<0.0040*
Constant As	81.061815	2.063234	39.29	<0.0001*
Constant Se	44.041606	4.826691	9.12	<0.0001*

*Significant at the 95% confidence level.

4 Discussion

The regression coefficient (R^2) is a number that indicates the proportion of the variance in the dependent variable that is predictable from the independent variable [15]. It is a statistic used in the context of statistical models whose main purpose is either the prediction of future outcomes or the testing of hypotheses, on the basis of other related information. It provides a measure of how well observed outcomes are replicated by the model, based on the proportion of total variation of outcomes explained by the model [16, 17].

The R^2 of Cd, Pb, Ni, Cr, As and Se were respectively 0.59; 0.80; 0.39; 0.81; 0.70 and 0.86. The coefficient of determination R^2 in this study were relatively high, indicating a good agreement between the model predicted and the experimental values.

The plots of the experimental value versus the predicted value for Cd, Pb, Ni, Cr, As and Se are shown in Fig. 1. Since the p-value for the model was lowers than 0.05, there was a statistical relationship between Cd, Pb, Ni, Cr, As and Se removal; and the selected variables at a 95% confidence level; as can be observed in the estimation of coefficients table (Table 4).

Figure 2 shows the three-dimensional response surface which was constructed to show the effects of the coagulant dosage and the pH on Cd, Pb, Ni, Cr, As and Se reduction from textile wastewater by coagulation flocculation processes with FeCl3. The flocculant concentration was fixed at 2.6 mL/L. The interaction between these factors (FeCl3 and pH) causes the most significant variation in Cd, Pb, Ni, Cr, As and Se removal as can be observed at Fig. 2, for example, if we set the coagulant dosage at 0.64 mg/l and pH 8.1; the removal of Cd, Pb, Ni, Cr, As and Se can achieve respectively; 40%, 78%, 62%, 22%, 81% and 44%.

These results show that the coagulation–flocculation mechanism differs depending on the pH value and dosage of coagulant and flocculant. Several studies have reported the examination of this process for the treatment of industrial wastewater, especially with respect to performance.

Optimization of coagulant/flocculant, determination of experimental conditions, assessment of pH and investigation of flocculant addition [18].

The goodness of fit of heavy metals removal was evaluated by the regression coefficient R^2 (between 0.86 and 0.59). The 86% sample variation observed for Se removal was attributed to the variables selected (pH, coagulant and flocculant dosages), while the model did not explain 14% of the variations. Another way to assess the goodness of fit of the model is by plotting the experimental values versus the predicted values [19]. As can be seen, the model approximately total represents the experimental data over the range studied. The plot shows the best fit as it may also be observed by the regression coefficient.

5 Conclusion

This study proved the effectiveness of the coagulation flocculation process with ferric chloride as a coagulant and polymer as a flocculant in the treatment and the removal of heavy metals from a real textile wastewater. The influence of three factors (coagulant dose, flocculant dose and pH) was investigated and modeled through response surface methodology and found to be significant and fundamental for maximizing the performance of the process. In addition, the optimization of the process based on RSM was employed to increase the removal of response variables and for seeking the optimal conditions.

The results show that the removal efficiency of heavy metals from real textile wastewater by coagulation–flocculation differs depending on the pH value and dosage of coagulant and flocculant. The optimal conditions obtained were 0.64 mg/l of coagulant, with 2.6 ml/l of flocculant at pH 8.1. The treatment under these conditions can remove 40% of Cd, 78% of Pb, 62% of Ni, 22% of Cr, 81% of As and 44% of Se.

Acknowledgment. Authors are thankful to Ministry of National Education, Vocational Training, Higher Education and Scientific Research (MNEVTHESR) (Morocco) for the financial support (This work is part of ERANETMED-Water Project "SETPROpER" project.).

References

1. Halimoon, N., Goh Soo, R.: Removal of Heavy Metals from Textile Wastewater using Zeolite. Environment Asia 3 (special issue), pp. 124–130 (2010)
2. Obiora-Okafo, I.A., Onukwuli, O.D.: Optimization of coagulation- flocculation process for colour removal from azo dye using natural polymers: response surface methodological approach. Niger. J. Technol. (NIJOTECH) **36**, 482–495 (2017)
3. Gosavi, V., Sharma, D. A general review on: various treatment methods for textile wastewater. J. Environ. Sci. Technol. 29–39 (2014)
4. Moi Pang, F., Ping Teng, Sh., Tow Teng, T., Omar, A.K.M.: Heavy metals removal by hydroxide precipitation and coagulation- flocculation methods from aqueous solutions. Water Qual. R. J. Canada **44**(2), 174–182 (2009)
5. Burke, G., Singh, B.R., Theodore, L.: Handbook of Environmental Management and Technology, 2nd edn., pp. 217–235. Wiley-Interscience, New York (2000)
6. Fan, M., Brown, R.C., Sung, S.W., Huang, C.P., Ong, S.K., Leeuwen, J.H.: Comparisons of polymeric and conventional coagulants in Arsenic(V) Removal. Water Environ. Res. **75**, 308–313 (2003)
7. Walsh, M.E., Zhao, N., Gora, S.L., Gagnon, G.A.: Effect of coagulation and flocculation conditions on water quality in an immersed ultrafiltration process. Environ. Technol. **30**, 927–938 (2009)
8. Ahmad, A.L., Wong, S.S., Teng, T.T., Zuhairi, A.: Optimization of coagulation–flocculation process for pulp and paper mill effluent by response surface methodological analysis. J. Hazard. Mater. **145**, 162–168 (2007)
9. Dominguez, J.R., Heredia, J.B., Gonzalez, T., Lavado, S.: Evaluation of ferric chloride as a coagulant for cork processing wastewaters. Influence of the operation conditions on the removal of organic matter and settleability parameters. Ind. Eng. Chem. Res. **44**, 6539–6548 (2005)
10. Fendril, I., Khannous, L., Timoumi, A., Gharsallah, N., Gdoura, R.: Optimization of coagulation-flocculation process for printing ink industrial wastewater treatment using response surface methodology. Afr. J. Biotechnol. **12**, 4819–4826 (2013)

11. Aziz, S.Q., Aziz, H.A., Yusoff, M.S., Bashir, M.J.K.: Landfill leachate treatment using powdered activated carbon augmented sequencing batch reactor (SBR) process: Optimization by response surface methodology. J. Hazard. Mater. **189**, 404–413 (2011)
12. Amuda, O.S., Amoo, I.A.: Coagulation/flocculation process and sludge conditioning in beverage industrial wastewater treatment. J. Hazard. Mater. **141**, 778–783 (2007)
13. Mufeed, S., Kafeel, A., Gauhar, M.R., Trivedi, C.: Municipal solid waste management in Indian cities. Waste Manag. **28**, 459–467 (2008)
14. Dikshit, A.K.: Treatment of landfill leachate using coagulation. In: 2nd International Conference on Environmental Science and Technology, IPCBEE 6. IACSIT Press, Singapore (2011)
15. Camba, A., González-García, S., Bala, A., Fullana-i-Palmer, P., Teresa Moreira, M., Feijoo, G.: Modeling the leachate flow and aggregated emissions from municipal waste landfills under life cycle thinking in the Oceanic region of the Iberian Peninsula. J. Clean. Prod. **67** (15), 98–106 (2014)
16. Ahmad, A.L., Ismail, S., Bhatia, S.: Optimization of coagulation– flocculation process of palm oil mill effluent using response surface methodology. Environ. Sci. Technol. **39**, 2828–2834 (2005)
17. Carvalho, G., Delée, W., Novais, J.M., Pinheiro, H.M.: A factorially- designed study of physico-chemical reactive dye color removal from simulated cotton textile processing wastewaters. Color. Technol. **118**, 215–219 (2002)
18. Bathia, S., Othman, Z., Ahmad, A.L.: Coagulation–flocculation process of POME treatment using Moringa oleifera seeds extract: optimization studies. Chem. Eng. J. **133**, 205–212 (2007)
19. Wang, J.P., Chen, Y.Z., Ge, X.W., Yu, H.Q.: Optimization of coagulation– flocculation process for a paper-recycling wastewater treatment using response. Colloids and Surfaces A: Physicochemical and Engineering Aspects. Issues 1–3, pp. 204–210 (2007)

Monitoring of the Purification Quality of the M'Rirt Station with a View to Its Extension and the Optimization of Its Performances

Imane El Alaoui[1](✉), Souad El Hajjaji[1], and Mohamed Saadlah[2]

[1] Laboratory of Spectroscopy, Molecular Modeling, Materials, Nanomaterials, Water and Environment, Chemistry Department, Faculty of Sciences, Mohammed V University, Rabat, Morocco
i.alaoui1994@gmail.com
[2] The Water Quality Control Department, National Office of Electricity and Drinking Water, Bouregreg Complex, Rabat, Morocco

Abstract. Over the past decade, the National Office of Electricity and Drinking Water (ONEE) has made liquid sanitation one of his strategic missions. Thus, a rigorous follow-up of the purification performances of these various stations is necessary in order to identify their failures and dysfunctions. Among these stations, there is the station of M'Rirt that is part of the Khénifra province, in the Beni Mellal-Khénifra region. The station is spread over an area of 9 ha and is located 3.5 km from M'rirt downtown. The M'Rirt wastewater treatment plant has been operational since June 2003 and has a capacity of 31,000 population equivalents. The treatment of wastewater is done by an extensive purification process, which is natural lagooning. It is designed mainly for the treatment of wastewater from the town of M'Rirt in order to preserve the Tighza wadi in which they are dumped. The M'Rirt station was designed to treat 1800 m³/d of wastewater and a BOD_5 pollutant load of around 1200 kg/d (2010 horizon). This natural lagoon station consists of 4 anaerobic basins, 4 optional basins, and 6 drying beds. The analysis of the purification performance of the M'Rirt station from 2013 to 2016 has shown that its purification yield is very low (27%). This state of fact is due on the one hand to the non-conformity of physico-chemical parameters, on the other hand to the hydraulic and the organic overload operation. To solve this malfunction, ONEE proposed transforming the optional lagoons into an aerated stage.

Keywords: Raw sewage · Natural lagooning · Purification performances

1 Introduction

Water is a precious good that undergoes various types of pollution and degradation: Ecosystems and the health of people are directly impacted [1, 2]. In 2025, nearly half of the population of the Mediterranean countries will be in a situation of tension or lack of water [3–6]. This shortage is accentuated by the deterioration of the quality of water resources under the influence of various sources of pollution (domestic, industrial,

© Springer Nature Switzerland AG 2019
M. Ezziyyani (Ed.): AI2SD 2018, AISC 913, pp. 267–279, 2019.
https://doi.org/10.1007/978-3-030-11881-5_24

agricultural wastewater, etc.) [7, 8]. Wastewater treatment has therefore been imposed to preserve the quality of natural environments, particularly surface and groundwater [9]. Purified water is currently largely rejected. For this reason, the reuse of these waters can help to fill part of the water deficit currently experienced by several countries, particularly in arid zones [10].

Wastewater treatment follows a logic of preservation of water resources and protection of the environment. They make it possible to greatly reduce the quantity of polluting substances contained in the wastewater so that it does not degrade the natural environment [11]. It is in this perspective of preservation of water resources and protection of the environment that the National Office of Electricity and Drinking Water (ONEE) decided, since 2001, to introduce a new dynamic to its intervention in the field of sanitation. This intervention is now part of the National Sanitation Program (NAP) [12]. This program targeted the installation of 330 wastewater treatment plants by 2020 [12]. Thus, it was necessary to think simultaneously of the different effluent management methods and products resulting from these stations, in this case: purified water, sludge and biogas.

The majority of purification systems are of the natural lagooning type (approximately 80% [13]). In fact, natural lagooning is simple to operate and less expensive than activated sludge purification systems or chemical treatment [14, 15]. Among the natural lagooning wastewater treatment plants, there is the resort town of M'Rirt which represents significant wastewater treated resources. These waters can be reused especially in agriculture provided that he has a mastery of problems related to public health and the environment.

The present study consists of analyzing the purification performance of the M'Rirt wastewater treatment plant with a view to its extension and optimization.

2   Materials and Methods

2.1   Location of the Study Site

The sewage treatment plant in the city of M'Rirt is part of the province of Khénifra, in the Beni Mellal-Khénifra region. The station is spread over an area of 9 ha and is located 3.5 km from the M'rirt downtown, or 400 m past the dwellings of the city on the left side of the main road connecting Meknes to Khénifra. The choice of the location of the station results from a study which relates to the field and which takes into account in the first place the remoteness of the houses, and secondly the topography which favors the gravity system between the basins of the station.

The M'Rirt wastewater treatment plant has been operational since June 2003 and has a capacity of 31,000 population equivalents. The treatment of wastewater is done by an extensive purification process, which is natural lagooning. It is designed mainly for the treatment of wastewater from the town of M'Rirt in order to preserve the Tighza wadi in which they are dumped. The latter is a temporary wadi which undergoes in the city a sudden change of direction: flowing from the South to the North, his bed abruptly takes the direction of the East at the center of M'Rirt.

2.2   Basic Data of the Station

Wastewater from the town of M'rirt transits to the station via a main outfall Ø800. The M'Rirt station was designed to treat 1800 m³/d of wastewater and a BOD₅ pollutant load of around 1200 kg/d (2010 horizon). This natural lagoon station consists of 4 anaerobic basins and 4 optional drying beds (see Fig. 1).

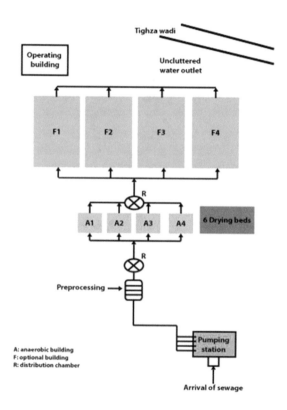

Fig. 1. Synoptic diagram of the M'Rirt wastewater treatment plant.

2.3   Description of the Operating Method of the Station

Pretreatment. Its main objective is to eliminate a large part of the decantable and colloidal elements contained in a raw urban effluent (coarse elements, greases, oils, etc.). It consists of three processes and is at the head of the treatment plant: screening, grit removal, and deoiling-degreasing. Apart from screening, grit removal and deoiling are not generalized and remain dependent on processing requirements. The M'Rirt station is equipped upstream of a lifting station consisting of a storm weir, a screen basket, a flowmeter and three water discharge pumps to the pre-treatment system with a fourth emergency pump. In addition, it is equipped with a pretreatment system, which includes a manual bar screen inclined bars with 3 cm spacing.

Primary Treatment. The pre-treated effluent joins a primary distributor, which ensures an even distribution between the anaerobic basins. The latter have almost no oxygen on the majority of the water column, and they reduce nearly 60% of the initial Biochemical oxygen demand (DOB) load. The M'Rirt station has four anaerobic basins, which operate in parallel with a capacity of approximately 3400 m^3 and a residence time of the order of 4 to 5 days. These basins are four meters deep and cover an area of approximately 3444 m^2.

Secondary Treatment. The effluent leaving the anaerobic basins joins a second distributor to the optional basins, whose objective of the treatment is the elimination of organic pollution, thus of the Biochemical oxygen demand (BOD_5) and the Chemical oxygen demand (COD). The M'Rirt station includes four optional pools in parallel with a capacity of approximately 18,600 m^3 and a residence time of 40 days. These basins have a depth of one and a half meters and an average surface area of 12,238 m^2.

A Drying Bed. The drying bed is a work on which sludge is applied when extracted from digesters or clarifiers. The drying is carried out by natural evaporation and by drainage. The station has a drying bed with six bins for the treatment of sludge with an area of 1250 m^3. The fresh sludge is pumped back to these beds by mobile pumps.

Parameters and Methods of Wastewater Analysis. All the methods used for the analysis of the monitoring and control parameters are standardized according to Moroccan or international standards in force. The Suspended solids (SS) were determined by the gravimetric method (NM-03-7-052- 1996). Biochemical oxygen demand (BOD5) was measured by OXITOP (NM 03-7- 056-1997). The chemical oxygen demand (COD) was measured by the colorimetric method (MA 315-COD 1.0 CEAEQ Québec Standards Method, 5220D, 22nd Edition 2012). Nitrogen and total phosphorus were determined by the continuous flow colorimetric method (Skalar Methods Catmr 475-424).

3   Results and Discussion

3.1   Study of the Characteristics of Incoming Raw Wastewater at the M'Rirt Station

In order to evaluate the characteristics of incoming raw wastewater at the M'Rirt station, the results of flow measurements, pollutant loadings, physicochemical parameters (SS, BOD_5, and COD) of the incoming wastewater during the period from 2013 to 2016 were considered.

Evaluation of the Origin of Raw Wastewater From the M'Rirt Station. For a better determination of the typology of incoming raw wastewater at the M'Rirt station, we have used the usual ratios of urban wastewater, namely: Ratio COD/BOD_5 and SS/BOD_5 (see Table 1).

Table 1. The usual ratios of urban wastewater.

Limits of the ratio	COD/BOD_5	SS/BOD_5
Lower bound of the ratio	2	1.2
Upper limit of the ratio	2.5	1.5

The analysis shows that the recorded values of COD/BOD_5 during the 2013–2016 period are between 0.5 (september-october 2014) and 2.7 (june 2016). All COD/BOD_5 ratios are below the lower limit ($COD/BOD_5 = 2$), with the exception of those recorded during the months of january and june 2016, which are above the upper limit. Nine values are above the lower limit of the usual range (see Fig. 2).

Quantitative analysis of these ratios shows that 47% of the COD/BOD_5 ratios are below the lower limit of the usual range. Thus, it can be concluded that the wastewater entering the M'Rirt station is of a domestic character and has an easily biodegradable organic load.

The values of the ratio SS/BOD_5 obtained are below the ranges in comparison with that of the usual ratio where they are between 1.2 and 1.5 (see Fig. 3). They are around an average of 0.75 with a minimum value of 0.4 measured in October 2014 and a maximum value of 1.1 measured in March 2014, July 2015 and in January, April 2016.

The analysis of these values shows that 100% of the SS/BOD_5 ratios are lower than the lower limit value of the usual range. These low SS/BOD_5 ratios are mainly due to the settling of the suspended material at the network level.

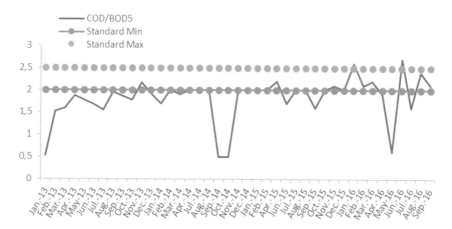

Fig. 2. Representation of COD/BOD_5 reports of wastewater from the M'Rirt station.

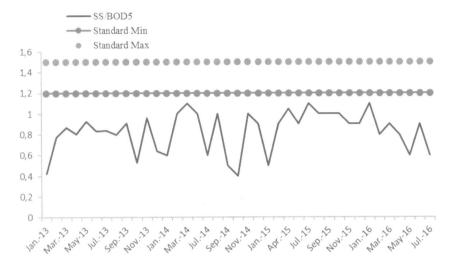

Fig. 3. Representation of SS/BOD₅ reports of wastewater from the M'Rirt station.

Quantitative Evaluation of Incoming Flow at the Station. The measured flow measurement results during the 2013–2016 period are between 2061 m³/d (February 2013) and 4500 m³/d (April 2015) (see Fig. 4). The comparative analysis of these results according to the nominal flow rate which is 1800 m³/d shows that the M'Rirt station operated in hydraulic overload ranging from 50 to 150% during this period.

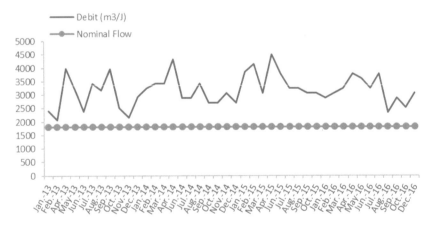

Fig. 4. The variation of the feed rate of the M'Rirt station.

Evaluation of the Incoming Pollutant Load at the M'Rirt Station. The pollutant load represents the amount of pollution transiting during a defined time, usually one day in the network. It is expressed in Kg BOD_5/d, Kg COD/d, Kg SS/d, and is determined from the following formula (1):

$$\text{Pollutant load} = [BOD_5(mg/L) * \text{Debit } (m^3/d)]/1000 \qquad (1)$$

The results of the evolution of the incoming pollutant load at the station show a doubling of the pollutant load over a number of months during the period from 2013 to 2016 (see Fig. 5). Indeed, the comparative analysis of incoming pollutant loads at the M'Rirt station with the nominal pollutant load (1200 kg/d), showed an overflow of 92%. It shows that the M'Rirt station operates in organic overload.

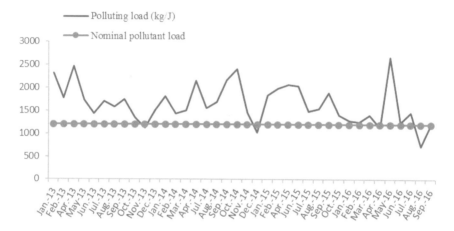

Fig. 5. Evolution of the incoming pollutant load at the M'Rirt station.

Qualitative Assessment of Effluents Entering the M'Rirt Station. To evaluate the quality of raw wastewater entering the station, the physico-chemical pollution parameters are compared: SS, BOD_5 and COD (see Figs. 6, 7 and 8) at the station entrance to the usual urban wastewater ranges in Morocco (see Table 2).

Table 2. Usual ranges of urban wastewater in Morocco.

Pollution parameters	Usual range (mg/L)
SS	250–500
BOD_5	200–400
COD	500–800

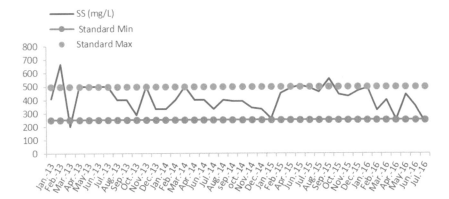

Fig. 6. Evolution of the SS at the entrance to the M'Rirt station.

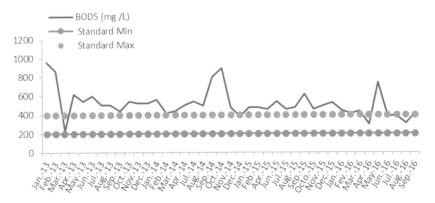

Fig. 7. Evolution of the BOD₅ at the entrance to the M'Rirt station.

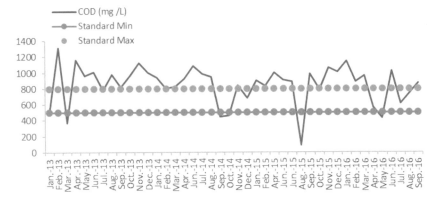

Fig. 8. Evolution of the COD at the entrance to the M'Rirt station.

The measurement results of the MES recorded during this period are included in the usual range, with the exception of the four values that are above the upper limit value recorded during the month of February 2013 and June, September 2015 and April 2016 (see Fig. 6).

The percentage of suspended solids measured at the entrance to the M'Rirt station, which is included in the usual range of urban wastewater, is 87%.

The concentrations of BOD_5 recorded during the period 2013–2016 are not included in the usual range of urban wastewater, except for six values that are below the upper limit value recorded in March 2013, in December 2014 and in April, June, July and August 2016 (see Fig. 7). The percentage of BOD_5 values is not included in the usual range of the order of 87%.

The results of the COD measurements recorded during the 2013-2016 period show that only 15% of the values are included in the usual range and 85% of the values are out of range of which 73% are above the upper limit and 12% below the lower limit (see Fig. 8).

3.2 Quality of Treated Wastewater at the Exit of M'Rirt Station

The SS, BOD_5, and COD values at the M'Rirt station output has been considered to study their level of compliance in comparison with the Moroccan standards for domestic discharge (see Figs. 9, 10 and 11).

In comparison with the specific liquid discharge limits, out of 46 samples analyzed during the period from 2013 to 2016, the effluent at the outlet of the station has values that do not comply with the specific limit values set for the three main pollution parameters (SS, DBO_5 and COD). These ones are of the order of 36% for the SS values, 80% for the BOD_5 values, and 85% for the COD values.

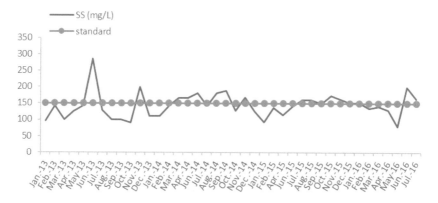

Fig. 9. Evolution of the SS at the exit of the station.

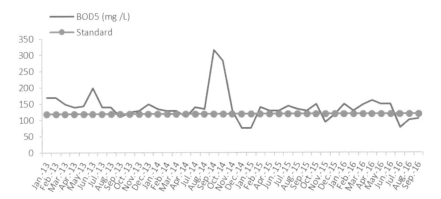

Fig. 10. Evolution of the BOD₅ at the exit of the station.

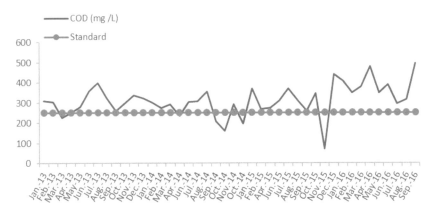

Fig. 11. Evolution of the COD at the exit of the station.

3.3   Analysis of the Purification Performance of the M'Rirt Station

The following figures show the rates of reduction in SS, BOD₅ and COD at the M'Rirt station and the treatment efficiency of total phosphorus and total nitrogen in the station (see Figs. 12, 13, 14 and 15).

The analysis of these physicochemical parameters shows that their performance is low in comparison to the European directive on the treatment of urban wastewater.

(Figure 15) illustrates, during the period 2013–2015, the purification performances for the nitrogen and phosphorus pollution, which are not reached at the M'rirt station. This can be justified by the treatment method used.

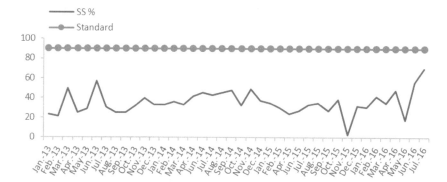

Fig. 12. Abatement of the SS at the M'Rirt station.

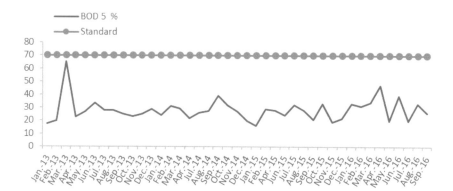

Fig. 13. Abatement of the BOD$_5$ at the M'Rirt station.

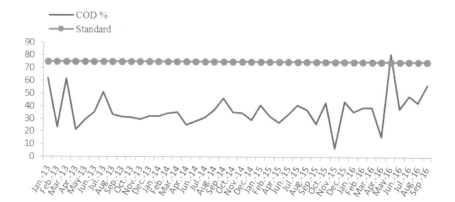

Fig. 14. Abatement of the COD at the M'Rirt station.

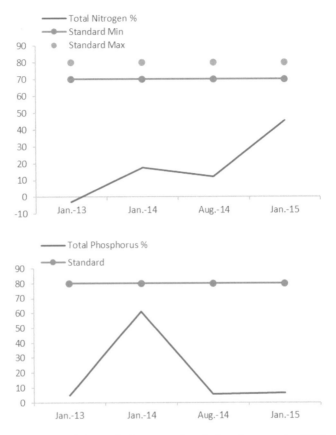

Fig. 15. Abatement of the total nitrogen and total phosphorus at the M'Rirt station.

4   Conclusion

Based on the results of the purification performance of the M'Rirt station during the 2013–2016 period for raw and untreated sewage, it can be concluded that:

The M'Rirt station operated in hydraulic overload up to 150% and in organic overload of 92%. The wastewater at the entrance of the M'Rirt station is charged (exceeding the prescribed value 1200 kg/d) and the effluent at the exit of the station is not compliant with Moroccan standards of domestic discharge for the three Physico-chemical parameters SS, DBO$_5$ and COD. For physico-chemical parameters, the abatement results are very low (27% of BOD$_5$). This is due to the hydraulic and organic overload and the algal proliferation knowing that the cleaning of the anaerobic basins was carried out in 2010. This dysfunction generates nauseating odors which constitutes a source of nuisance for the population and can be considered as a danger on the natural environment (oued Tighza). According to these results, it becomes crucial to intervene to remedy this critical situation. Thus we suggest to transform the optional lagoons into aerated stage: This solution is the most advantageous as it allows to support the hydraulic and organic loads by 2030 with modifications at the level of the existing basins (increase of the volumes by overcrowding or elevation) and eliminate odors.

Acknowledgement. The authors would like to thank the Laboratory of Spectroscopy, Molecular Modeling, Materials, Nanomaterials, Water and Environment, (LS3MN2E-CERNE2D), Mohammed V University; The Water Quality Control Department (DCE), of the National Office for Electricity and Potable Water (ONEE) for their technical support of this research.

References

1. Schwarzenbach, P.R., Egli, T., Hofstetter, B., Gunten, U., Wehrli, B.: Global water pollution and human health. Annu. Rev. Environ. Resour. **35**, 109–1369 (2010)
2. Schwartzbrod, L., Pommepuy, M., Albert, M., Dupray, E., Boher, S., Beril, C., Guichaoua, C.: Influence of sewage discharge on the pollution of an estuary (Morlaix, France). Zentralbl. Hyg. Umweltmed. **192**(3), 238–247 (1991)
3. Vörösmarty, J.C., McIntyre, P.B., Gessner, M.O., Dudgeon, D., Prusevich, A., Green, P., Glidden, S., Bunn, S.E., Sullivan, C.A., Reidy Liermann, C., Davies, P.M.: Global threats to human water security and river biodiversity. Nature **467**, 555–561 (2010)
4. Yang, H., Zehnder, A.J.B.: Water scarcity and food import: a case study for southern Mediterranean countries. World Dev. **30**(8), 1413–1430 (2002)
5. Iglesias, A., Garrote, L., Flores, F., Moneo, M.: Challenges to manage the risk of water scarcity and climate change in the mediterranean. Water Resour. Manag. **21**(5), 775–788 (2007)
6. Laraus, J.: The problems of sustainable water use in the mediterranean and research requirements for agriculture. Ann. Appl. Biol. **144**(3), 259–272 (2004)
7. Falkenmark, M., Widstrand, C.: Population and water resources: a delicate balance. Popul. Bull. **47**(3), 1–36 (1992)
8. Peters, E., Meybeck, M.: Water quality degradation effects on freshwater availability: impacts of human activities. Water Int. **25**(2), 185–193 (2000)
9. Guessab, M., Bize, J., Schwartzbrod, J., Mault, A., Morlot, M., Nivault, N., Schwartzbrod, L.: Waste water treatment by infiltration-percolation on sand: results in Ben Sergao, Morocco. Water Sci. Technol. **27**(9), 91–95 (1993)
10. Al jayyousi, O.R.: Greywater reuse: towards sustainable water management. Desalination **156**(1-3), 181–192 (2003)
11. Aziz, N.A., Salih, S.M., Hama-Salh, N.Y.: Pollution of tanjero river by some heavy metals generated from sewage wastewater and industrial wastewater in sulaimani district. J. Kirkuk Univ. Sci. Stud. **7**(1), 67–84 (2012)
12. http://www.iea.ma/en/actualites/365-pna-seminaire-national-sur-l-assainissement-liquide-15-16-mars-2017-a-marrakech. Last Accessed 10 Apr 2018
13. Racault, Y. (Cemagref): Le Lagunage naturel: les leçons tirées de 15 ans de pratique en France. 1ére Edn. (1997)
14. Shuval, H.I., Vekutie, P., Fattah, B.: An epidemiological model of the potential health risk associated with various pathogens in water irrigation. Water Sci. Technol. **18**(10), 191–198 (1986)
15. Fagrouch, A., Amyay, S., Berrahou, A., El halouani, H., Abdelmoumen, H.: Performances d'abattement des germes pathogènes en lagunage naturel sous climat aride: cas de la filière de traitement des eaux usées de la ville de Taourirt. Afrique Sci. **6**(3), 87–102 (2010)

Impact of the Mohammedia Benslimane Landfill on the Quality of Groundwater and Surface Water

Hamri Zineb[✉], Mouhir Latifa, Saafadi Laila, and Souabi Salah

Faculty of Sciences and Techniques of Mohammedia,
Hassan II University, Casablanca, Morocco
hamri.zineb.1992@gmail.com, latmouh@gmail.com,
laila.saafadi@gmail.com, salah.souabi@gmail.com

Abstract. This work aims to assess the environmental impact of the Mohammedia-Bensliman landfill on groundwater resources. As a result, a companion sampling of the groundwater and surface water of Ouad Nfifikh was carried out around the landfill. In addition, characterization of the pollutant load of raw leachate was also performed to better identify the potential hazard of release to the environment, in which excessive levels of chloride, BOD5, COD, nitrate, ammonium and phosphorus are present. The value of the BDO5/COD ratio which is of the order of 0.5 indicates pollution by the biodegradable materials. The pH value and the high content of bicarbonates suggest the prevalence of the acidogenic phase, which corresponds to the age of the rejection (<10 years), the high concentration of the main ions obtained (chloride, total hardness, bicarbonate) increases the Mineral content Heavy metals in the analyzed waters represent low concentrations compared to imposed standards.

Keywords: Contamination · Groundwater · Surface water · Leatchate

1 Introduction

Landfilling is the method of treatment adopted in Morocco. This is the management method currently used which consists of bulk storage of waste collected without prior treatment, the nature and large quantities produced represent real challenges in terms of management and treatment.

In the context of the protection and resolution of environmental problems, Morocco has placed great importance on good waste management strategies, considering the risks of contamination of groundwater and surface water by infiltration of leachates; which result from the percolation of rainwater through the deposits of waste added to the quantity of liquids contained in the waste itself and those resulting from their degradation; through the water table, this is the case of several countries other than Morocco such as Ukraine, Negiria, Egypt, China, Algeria … [1–5].

The landfill of Mohammedia-Benslimane is located in a rural agricultural area, surrounded by cultured fields, wells with depths ranging from 60 to 90 m, and Ouad Nfifikh which joins the seawater of the city of Mohammedia to its extremity.

© Springer Nature Switzerland AG 2019
M. Ezziyyani (Ed.): AI2SD 2018, AISC 913, pp. 280–288, 2019.
https://doi.org/10.1007/978-3-030-11881-5_25

The analysis of the leachates generated by this discharge showed it is a percolate with a high mineral and organic load (chlorides: 6957, 65 mg/l, COD: 37267, 5 mg/l, DBO5: 19000 mg/l and nitrate: 55,74 mg/l).

The physico-chemical study of well water in the region does not show a significant deterioration in the quality of the water analyzed. Indeed, electrical conductivities (in average of 2 mS/cm) highly correlated with chloride (790 mg/l), This indicates that there is no significant contamination of the water of these wells by leachates. What is the organic and metallic charge (practically negligible in natural groundwater), low levels of chemical and biological oxygen demand.

Study Site

The public dump of Mohammedia-Benslimane has started its contract since 2012 for a period of 20 years of service. This is why the old Mohammedia dump was closed due to saturation. During the rainy season, leachate can flow into Wadi Nfifikh and infiltrate the soil or evaporate in the air. The landfill is located about 7 km from the center of ben yekhlef. The area of the landfill covers 47 ha and concerns the following municipalities: Mohammedia, Beni Yakhlef, Echellalate, Benslimane, Bouznika and Mansouria, with a population of 409,000 inhabitants.

There are several types of materials such as: paper and cardboard, plastics, glass, metals (cans and paint), bones, textiles, leather … This diversity of waste generates leachates rich in organic and mineral matter, which can be to fear the great contamination of groundwater and surface water.

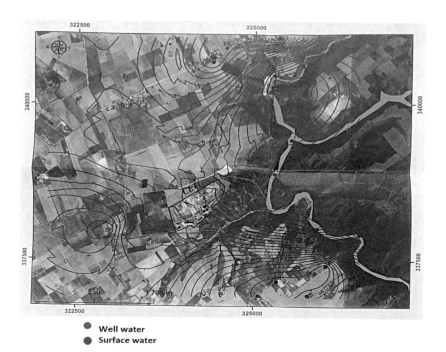

● Well water
● Surface water

Fig. 1. Distribution of well water withdrawal points and surface water

Sampling and Analysis

We chose the sampling points so that the sample is representative of the environment according to its position relative to the geographical location of the landfill. Figure 1 shows the distribution of the different sampling points chosen for the study. The sampling was divided into two types: 9 groundwater points (P1–P9 wells) and 3 surface water points (Ouad Nfifikh: P_{amant}-P_{aval}-P_{chaaba}).

The main physicochemical parameters analyzed for leachate and water are: turbidity (turb), electrical conductivity (EC), dissolved oxygen (dissolved O2), pH, chloride concentrations (Cl-), total hardness TH, bicarbonates TA and TAC, chemical oxygen demand (COD), nitrite and nitrate and ammonium were determined by volumetry [6]. Biological oxygen demand (BOD5) was measured by a BOD meter. Total nitrogen (TN) was estimated using the Kjeldhal method.

Heavy metals were analyzed for spring waters (iron, zinc, chromium, cadmium, copper) were measured using an atomic absorption spectrophotometer.

2   Results and Discussion

Physical and Chemical Characteristics of Leachate

The results of the physical and chemical analyzes of leachate are presented in Table 1. This table shows that the pH is 8.29, which is suitable for the methanogenic phase corresponding to the production of methane and carbon dioxide [7]. Similar results were found by [8] who found that the samples had a slightly higher pH and did not exceed the value of 8.84. On the other hand, [9] found that the average pH was 7.4 for municipal releases in India.

Table 1. Physical and chemical characteristics of leachate

Parameters	Unit	Leachate sample
pH		8,29
EC	µs/cm	4070
COD	mg/l	31720
DBO5	mg/l	10000
TSS	mg/l	4300
NO3-	mg/l	19,8
NH4+	mg/l	3300
TKN	mg/l	4775,99
P	mg/l	12
Cl-	mg/l	7090,6
DBO5/COD		0,32

[10] assessed leachate leachate characteristics and found average conductivity values and chlorides were 4257 µs/cm and 9683 mg/l respectively. This finding confirms the results of the present study where the conductivity is 4070 µs/cm with and the

chloride value is 7090.6 mg/l. Lower results were obtained [11] which found that the leachate conductivity of the Liosia landfill, located in the Attica region of Greece, was 24,038 μs/cm and the chlorides was 4149 mg/l. Although (El-salam and Abu-zuid [2]) showed that leachates collected in the egypt landfill have a conductivity and high chlorides of 40921 μs/cm 11387 mg/l respectively.

In the present study, the BOD5 has a value of 10000 mg/l and the COD is 31720 mg/l, the ratio BOD5/COD = (0.32) indicates that the leachates taken is of intermediate type [7] and tend to stabilize, it is the same case for the discharge of India which has a BOD5 of 19000 mg/l and the COD of 27200 mg/l [12]. [8] studied the leachate of the El Jadida municipal dump in Morocco and found that leachate had average COD and BOD5 values of 950 mg/l and 55.33 mg/l respectivel, BOD5 and COD were 0.06, these values indicate that the leachate was stabilized and that it was in the ripening phase, the same results were approved by [2]. Lower results were recorded in another study in Greece [13] comparing two landfills, the first with age <2 years with the leachate COD value ranging from 3000–60,000 mg/l and the 2000 BOD5 to 30,000 mg/l, and the other >10 years the leachate COD value ranging from 100–500 mg/l and the BOD5 from 100 to 200 mg/l.

The young leachates are more polluted than the mature ones where the COD can reach 72 000 mg/l for the young cases of the discharge of Agadir [14] and the 920 mg/l for the mature samples case of the discharge of el jadida [8]. The BOD5/COD ratio in young discharges where the biological activity corresponds to the acitogenic phase of anaerobic degradation that is greater than 0.5, the old landfills produce stabilized leachates with a relatively low COD and low biodegradability (ratio BOD5/COD < 0.1) [7]. In the present study, the variation of the different parameters can be influenced by several factors such; the type of waste and its characteristics, the absence of shredding before disposal, the compaction of waste that delays degradation, and meteorological conditions such as temperature and pressure.

The concentrations of kjeldhal nitrogen (NTK), ammonium (NH4+), nitrates (NO3-) and total phosphorus (P) for the leachate sample in our study are respectively 4775.99 mg/l; 3300 mg/l; 19.8 mg/l; 12 mg/l. Lower values were observed by [2] which gave as value of nitrates 1.4 mg/l and phosphors 0.37 mg/l. On the other hand [9] gave high values where the nitrates are of 22.36 mg/l and the phosphors of 22.29 mg/l. The high nitrate values indicate that the environment was oxidized and through the denitrification phenomenon, the nitrate ions are transformed into molecular nitrogen during the deacon reaction time [15]. The presence of the NTK is due to the phenomenon of mineralization which is a process of transformation of organic compounds into mineral compounds [16] so a part of the organic nitrogen which has turned into mineral nitrogen.

Assessment of the Impact of Leachate on Groundwater

The Mohemmadia-Benslimane landfill site is impervious due to the nature of the geology of the region, which is formed by limestones, basalts, siltutes and lower clays, siltstones and conglometric sandstones and a Psammitic basement [17]. The leachate generated do not show an infiltration announced through the subsoil to reach the underlying water table circulating at very great depths (25 to 140 m). The flow direction of the water table is directed in the direction of flow of Ouad Nfifikh towards the ocean in the north-east direction of the main entrance of the landfill, which can lead to the propagation of the pollution flow to Ouad Nfifikh.

Physical and Chemical Characteristics of Well Water

According to Table 2, all the wells monitored during this study have moderately large electrical conductivities ranging from 1092 to 3850 µs/cm, according to the Moroccan standards of water potability accept as an average value of electrical conductivity a value of 1300 µs/cm [18], in our case it excesses values that fall within the range and other higher standards. [19] found that the mean value of the conductivity does not exceed 1900 µs/cm, in [5] it has been found that the electrical conductivities in the study area vary from a value below 450 µs/cm to a value greater than 3000 µs/cm. To some extent, turbidity can reflect the degree of contamination, [4] found that the electrical conductivity values (4000 µs/cm) indicate presence of either contamination by leachate. On the other hand, [2] have recorded values that do not exceed 21.50 µs/cm. The salinity of well water may be due to the nature of the geology of the area. The pH of the samples studied is a basic pH and meets the standards of potability, the bicarbonate ions strongly exert in the waters of the wells since the $4.35 < pH < 8.34$ [20].

The results obtained show that chloride contents vary from 1226.7 to 411.25 mg/l of P1 to P7, these values largely exceed the standards of potability (300 mg/l) whereas P8 and P9 comply with the standards of Moroccan potability, [8] showed that the chloride content is very important with an annual average of 1050 mg/l, this high load of chloride ions is probably related to an accidental contamination by leachate that are heavily loaded with chloride (7090.6 mg/l) or it is related to the effect of the nature of geology. The chloride ion is a very mobile element, which migrates easily to the underlying aquifers. It is not affected by the adsorption or ion exchange phenomena, it

Table 2. Physical and chemical characteristics of well water

	Unit	P1	P2	P3	P4	P5	P6	P7	P8	P9
TURB	NTU	0.32	0.32	3.99	1.95	<0.01	0.15	19.2	0.78	<0.01
EC	µs/cm	3640	1686	2940	2390	3230	3730	3850	1092	1524
pH		7.391	7.527	7.229	7.254	7.003	7.481	7.403	8.031	7.155
O2	mg/l	7.53	9.03	7.58	9.20	8.77	8.91	6.40	8.25	7.20
NTK	mg/l	0.77	<0.1	<0.1	<0.1	<0.1	112.45	<0.1	<0.1	2.89
NH4+	mg/l	<0.01	<0.01	<0.01	<0.01	<0.01	0.073	0.075	<0.01	<0.01
NO3-	mg/l	6.5	3.6	1.4	2.5	4.1	16.3	26.5	4	5.7
NO2-	mg/l	0,03	0.026	0.054	0.018	0.023	0.019	0.052	0.025	0.02
TH	mg/l	860	400	400	600	640	680	720	380	360
TA	mg/l	0	16	0	0	0	0	0	20	0
TAC	mg/l	300	160	280	240	260	250	260	200	360
CL-	mg/l	1226,7	411,25	794,15	659,43	1524,5	872,14	815,42	177,27	226,9
DCO	mg/l	62	47	48	30	40	41	36	40	41
DBO5	mg/l	<1	<1	5	<1	<1	<1	1	<1	1
P	mg/l	<0.5	<0.5	<0.5	<0.5	<0.5	<0.5	<0.5	<0.5	<0.5

does not intervene in the acid-base or oxidation-reduction reactions and it is not retained by the complexes clay-humic soils. This is why it is frequently used as a good conservative tracer that can highlight the impact of leachate on the physicochemical quality of the groundwater [8, 11].

Natural groundwater generally contains little or no organic matter [8]. However, despite the proximity of the dump wells (a radius of 4 m), they have very low COD contents, with averages of 43 mg/l with maximum of 62 mg/l these values can be explained by the great depth of well water which creates a deficiency in O_2 or a contamination of the leachate. [12] found that the chemical oxygen demand varies between 2 and 17 mg/l, whereas [11] found high values which can reach up to 90 mg/l. It is a contamination by the runoff that is affected leachate. The values of the BOD5 are negligible in our case study.

Physical and Chemical Characteristics of Surface Water

According to Table 3, the concentrations of biodegradable organic matter (BOD5) are negligible in surface water in our study (Ouad Nfifikh) with a somewhat high value in P_{chaaba} which links the dump and the river, [19] found a lower value than ours (DBO5 = 18.85 mg/l). The concentrations of oxidizable organic matter (COD) are also a little important especially in the P_{chaaba} which reaches a value of 1537 mg/l, it means a direct contamination of the leachate by streaming rain water through the discharge.

Table 3. Physical and chemical characteristics of surface water

	Unit	P_{amant}	P_{aval}	P_{chaaba}
TURB	NTU	384	58.4	6.24
EC	µs/cm	471	1320	5.95
pH		7.595	7.579	7.868
O2	mg/l	9.47	9.14	2.55
NTK	mg/l	12.73	3,92	18.08
NH4+	mg/l	0.266	1.466	2.78
NO3-	mg/l	2.7	2.5	2.4
NO2-	mg/l	0.071	0.294	0.214
TH	mg/l	180	340	1500
TA	mg/l	0	0	30
TAC	mg/l	120	160	450
CL-	mg/l	99,268	290,71	1545,8
DCO	mg/l	59	56	1537
DBO5	mg/l	<1	3	30
P	mg/l	<0.5	<0.5	<0.5

Recorded electrical conductivity values ranged from 5.95 to 1320 µs/cm. Just as the pH values are basic, so a dominance of bicarbonate ions is the same for [21].

The results of the analyzes of the samples show high concentrations of chlorides up to 1545.8 mg/l, a very low value of dissolved oxygen was recorded in the P_{chaaba} that confirms the contamination of the water of the Wad by the leachate.

Heavy Metals

Table 4 shows the major elements of heavy metals found Fe, Zn, Cr, Cd and Cu. The values observed in our study are negligible, and do not present a contamination of clear water by the leachate of the Mohammedia-benslimane landfill, Fe, Zn, Cr have respectively as maximum values 0.93 ppm; 0.58 ppm and 0.057 ppm, with absence of cadmium and copper metals. For [22] found lower values such as iron = 0.65 ppm; zinc = 0.08 ppm and Cr = 0.08 ppm.

Table 4. Heavy metals of surface water

	Unit	P1	P2	P3	P4	P5	P6	P7	P8	P9	P_{amant}	P_{aval}	P_{chaaba}
Fe	ppm	0.10	0.05	0.30	0.08	0.12	0.09	0.19	0.04	0.10	0.93	0.13	0.15
Zn	ppm	0.01	0.06	0.05	0.03	0	0.04	0.4	0.05	0.58	0.06	0	0.04
Cr	ppm	0.054	0.037	0.036	0.046	0.034	0.057	0.049	0.051	0.048	0.044	0.043	0.039
Cd	ppm	0	0	0	0	0	0	0	0	0	0	0	0
Cu	ppm	0	0	0	0	0	0	0	0	0	0	0	0

3   Conclusion

The public dump of Mohammedia-Benslimane does not show a strong contamination of well water and surface water studied according to the results of physicochemical analyzes and heavy metals.

The Mohammedia-Benslimane landfill is of controlled type because of the permeability of the site, the landfill has a fairly impermeable geology and a fairly high water table depth as well as the presence of drainage network and the recovery and recovery device. Leachate treatment with a well-studied waterproofing. Leachate analyzes show waters with high biodegradable content and high concentrations of COD, BOD5, nitrate, total phosphorus … which exceed the standards for direct discharges into the natural environment. Leachate is a source of contamination for its environment, especially for groundwater and surface water.

The assessment of the quality of the water table, through well water analyzes, shows the presence of certain pollutants present in the leachates analyzed. Concentrations are noted in certain elements: COD, chlorides, electrical conductivity which is not, their presence in clear water is not necessarily due to contamination by leachate, we can also note the nature of the geology that has a direct influence on the composition of the waters.

References

1. Makarenko, N., Budak, O.: Waste management in Ukraine: municipal solid waste land fills and their impact on rural areas. Ann. Agrar. Sci. **15**(1), 80–87 (2017)
2. El-salam, M.M.A., Abu-zuid, G.I.: Impact of landfill leachate on the groundwater quality: a case study in Egypt. J. Adv. Res. **6**(4), 579–586 (2015)
3. Akinbile, C.O.: Environmental impact of landfill on groundwater quality and agricultural soils in Nigeria, vol. 2012, no. 1, pp. 18–26 (2012)
4. Han, D., Tong, X., Currell, M.J., Cao, G., Jin, M., Tong, C.: Evaluation of the impact of an uncontrolled land fill on surrounding groundwater quality, Zhoukou, China. J. Geochem. Explor. **136**, 24–39 (2014)
5. Kehila, Y., Mezouari, F., Brahim, B., Matejka, G., Harrach, E.: Impact des décharges publiques sur l'environnement en Algérie: Quelles perspectives pour une gestion efficiente des lixiviats? (2001)
6. Rodier, J., Legube, B., Merlet, N.: Analyse de l'eau Rodier, 9ème edn, p. 1579 (2009)
7. Trabelsi, S.: Études de traitement des lixiviats des déchets urbains par les Procédés d'Oxydation Avancée photochimiques et électrochimiques. Application aux lixiviats de la décharge tunisienne 'Jebel Chakir,' p. 225 (2012)
8. Chofqi, A.: Environmental impact of an urban landfill on a coastal aquifer (El Jadida, Morocco), no. August (2015)
9. Naveen, B.P., Sivapullaiah, P.V., Sitharam, T.G., Ramachandra, T.V.: LAKE 2014: Conference on Conservation and Characterization of Leachate from Municipal Landfill and its Effect on Surrounding Water Bodies, no. November (2014)
10. El Kharmouz, M., Sbaa, M., Chafi, A., Saadi, S.: L'etude de l'impact des lixiviats de l'ancienne decharge publique de la ville d'oujda (maroc oriental) sur la qualite physicochimique des eaux souterraines et superficielles, pp. 105–119 (2013)
11. Fatta, D., Papadopoulos, A., Loizidou, M.: A study on the landfill leachate and its impact on the groundwater quality of the greater area, no. September (2016)
12. Mor, S., Ravindra, K., Dahiya, R.P., Chandra, A.: Leachate characterization and assessment of groundwater pollution near municipal solid waste landfill site, pp. 1–31
13. Papadopoulou, M.P., Karatzas, G.P.: Numerical modelling of the environmental impact of landfill leachate leakage on groundwater quality – a field application, pp. 43–54 (2007)
14. Jirou, Y., Harrouni, M.C., Belattar, M., Fatmi, M., Daoud, S.: Traitement des lixiviats de la décharge contrôlée du Grand Agadir par aération intensive. Rev. Marocaine des Sci. Agron. Vétérinaires **2**(2), 59–69 (2014)
15. Maton, S.R.D., Amalric, L., Ghestem, J.-Ph., Guigues, N.: Synthèse des cas de dénitrification naturelle dans les eaux souterraines en france: intérêt du processus pour restaurer la qualité des eaux (2000)
16. Limoges, D.: Voies de réduction des oxydes d'azote lors de leur injection dans un massif de déchets ménagers et assimilés (2005)
17. Farki, K., et al.: Mines et carrières triasico-liasiques de la région de Mohammedia : Inventaire, valorisation et étude d'impact environnemental [The triassic and liassic mines and quarries of Mohammedia area: inventory, valuation and environmental impact]. Int. J. Innov. Sci. Res. **20**(2), 306–326 (2016)
18. Benchekroun, R.H.: Normes de Qualité (2007)
19. Maiti, S.K., Hazra, T., Dutta, A.: Characterization of leachate and its impact on surface and groundwater quality of a closed dumpsite - a case study at Dhapa, Kolkata, India. Procedia Environ. Sci. **35**, 391–399 (2016)

20. Bourrie, G.: Relations entre le pH, l'alcalinité, le pouvoir tampon et les équilibres de CO2 dans les eaux nature Guilhem Bourrie To cite this version: HAL Id: hal-01189766 (2015)
21. Fournier, J., Lacarrière, B., Kechiched, R., Hani, A.: Assessing the feasibility of using the heat demand-outdoor Pina temperature function for a Haied district heat demand forecast. Energy Procedia **119**, 393–406 (2017). Water pollution and risk on of and Wadi Zied plain aquifer caused by the leachates of Annaba landfill (N-E Algeria)
22. Bouzid, S., Bouzid, S., Khannous, S., Zouag, M.A.: Contamination des eaux souterraines par le lixiviat des décharges publiques: Cas de la nappe phréatique R'Mel (Province de Larache - Maroc Nord-Occidental), vol. 5, no. June, pp. 1118–1134 (2011)

Author Index

© Springer Nature Switzerland AG 2019
M. Ezziyyani (Ed.): AI2SD 2018, AISC 913, pp. 289–290, 2019.
https://doi.org/10.1007/978-3-030-11881-5

Printed in the United States
By Bookmasters